THE UNEMPLOYMENT HANDBOOK

by Guy Dauncey

I am born to win!
I am born to win!
I've been rejected, considered dead,
I've been accused and been refused.
But the stone that the Builder refused
shall be the Head Cornerstone.
I cannot fail, 'cause I'm not alone.
I am born to win!

(*Jimmy Cliff*)

This Handbook is dedicated to YOU, the reader,
in the hope that together, we can win.

NATIONAL EXTENSION COLLEGE

A note about this handbook

This Handbook cannot be perfect. Nor can it please everyone. There are so many different unemployed people, and so many different attitudes that it is possible to take about unemployment. You will be sure to find places where you disagree with some of the things said, and other places where you think that it is not saying things that should be said.

The overall attitude that the Handbook takes is positive. It asks 'How can I make the most of this situation? Can I, or we, turn it into an opportunity?' It aims to help people to do this by offering ideas, information, advice, suggestions and encouragement. Any feedback, and suggestions for a future edition will be greatly welcomed by the author, if you write to him c/o the National Extension College, 18 Brooklands Avenue, Cambridge CB2 2HN.

The author

Guy Dauncey, the author, lives just outside the small market town of Ashburton on the edge of Dartmoor in south Devon, where he enjoys life greatly. He was unemployed himself for two years until he started making a living by writing about it. This Handbook started life in 1976 as a supplement to *Sherrack*, a local community magazine; it then became (for three years) a handbook for volunteers involved in 'Just the Job', a project for young unemployed people in the South-West run by the National Extension College. The author is considering a companion volume, an extension of Chapter 12, about the kind of changes that we need to be making in order to overcome unemployment and rebuild our economies.

He enjoys singing with the Ashburton Singers, loves music of all shapes and sizes, and is editor of *Interchange*, a small magazine concerned with personal change, the evolution of consciousness and the ideas of Teilhard de Chardin. As well as writing, he gardens and runs groups and workshops for unemployed people.

Acknowledgements

I would like to thank the following people, who have helped in various ways during the preparation of this Handbook: Barry Reeves, Barry Deller, Paddy Hall and everyone at 'Just the Job'; Hilary Eadson, Pat Kitto, Marion Warr, Janet Allbeson, Russell Sharp, Dave Hocking, N. A. Poolman, Andrew Pates, Martin Good; many personal friends; and John Meed, my editor, for his patience and support.

Finally, I mustn't finish without mentioning Joan Armatrading, Rod Stewart, Tomita, Kate and Anna McGarrigle, John Denver, Nina Simone, Holly Near*, Chris Williamson*, John Lennon, Franz Schubert and J. S. Bach, who kept me in such good company while the work slowly progressed.

Also to Jimmy Cliff and PPX Publishing for the quotation on the title page.

Guy Dauncey

*Holly Near's and Chris Williamson's music is available through WRPM, 62 Woodstock Road, Birmingham B13 9BN.

CONTENTS

HOW TO USE THIS HANDBOOK

Method 1: use it to find yourself answers to particular problems, such as improving your letter-writing, tackling your benefits problems, finding out about training and education, developing better interview techniques, etc. See the Contents, Index and the Problem Solver on p. vi.

Method 2: take a chapter that interests you, and work right through it from the beginning, doing the exercises and following up any suggestions.

Method 3: read Chapters 2 and 3 and do all the exercises, giving yourself plenty of space and time. These will help you to gain a better hold on your own situation, and to tackle it positively. Then use the other chapters as you need to.

Method 4: the Handbook lends itself well to use by a group. Work through the exercises with a friend, or with a self-help group (see Chapter 5). Help each other to do the exercises, to apply for rent rebates, to write better application letters, to tackle benefits problems, etc.using the Handbook as a support.

If you have any difficulty using this Handbook, then ask for some help. Some of the exercises are much better done with someone else, anyway. To find help:
- Ask a friend if she or he would give you a hand.
- Advertise locally, to make contact with other unemployed people (see Chapter 5).
- Contact the Citizens Advice Bureau, or another local advice centre (see (p. viii), and ask if there is someone who might help you.

YOU'VE GOT PROBLEMS? WE'VE GOT ANSWERS

For other problems, use the Contents page, and the Index.

LOCAL ADDRESSES AND PHONE NUMBERS

It will help you to use this Handbook if you fill in this page, using the telephone directory:

Address   *Phone No.*

Jobcentre Employment Office:
(Under 'Employment Services
Division' or 'Jobcentres')

Unemployment Benefit Office:
(Under 'Employment, Department of')

**Department of Health and Social
Security:**
(Under 'Health and Social Security,
Dept. of')

Careers Office :
(Under name of City or County
Council, then 'Careers Service', or
under 'Careers Service')

Social Services :
(Under name of County or City
Council then under 'Social
Services Dept.')

Youth Service* :
(Under 'Youth Officers', or under
name of City or County Council,
and then 'Youth Service' or 'Youth
Officers')

Employment Agencies:
(Yellow Pages)

Area Education Office:
(Under name of City or County
Council, then 'Education
Department')

Adult Education Department*:
(Try under name of City or
County Council, then 'Education Dept',
or 'Further Education')

(viii)

Address   *Phone No.*

Further Education Adviser*:
(Try under name of City or
County Council, then 'Education
Dept', or 'Further Education')

**Local College of Further Education,
or Technical College*:**
(as above, or under 'Schools and
Colleges' in the Yellow Pages)

'TOPS' Skillcentre:
(Under 'Training Services Division')

**Council for Voluntary Service or
Volunteer Bureau*:**

Library :
(Under name of City or County
Council, then 'Library Services')

Citizens Advice Bureau:

*You may find these hard to locate in the telephone directory. If you do, try phoning the Citizens Advice Bureau, and asking them for the information, or phone Directory Enquiries (192).

Chapter 1: The two crises

CRISIS 1: NATIONAL – CRISIS 2: PERSONAL

The national unemployment figure for the UK passed two million in August 1980 for the first time since 1935. 'Falling orders', 'Overseas competition', 'New technology', 'Streamlining', 'Rationalisation', 'Merger', 'Financial losses' – daily these slogans are flashed across the headlines as firms close and workers are sacked.

If we could honestly predict that these figures would fall at the end of the current recession, they could just be seen as a 'temporary setback'. But many studies point to even greater numbers out of work, recession or no recession, and there is talk of a 'new industrial revolution' with declining jobs forced on us by microelectronic computers running the workplace.

So we are facing a crisis; or rather, two crises. A national crisis – millions out of work; and a personal crisis – for the individuals who make up these horrific figures. The reason for writing this Handbook is that each crisis has two faces – one that just screams 'danger!' and one that says more quietly, 'opportunity'.

In Chinese, the word for 'crisis' is *wei-chi*, where *wei* actually means 'danger', and where *chi* means 'opportunity', or 'that from which all change comes'. The overall philosophy of this Handbook is therefore 'the *wei-chi* of unemployment'.

CRISIS ONE: NATIONAL

Danger

Five million unemployed? Gangs of angry, frustrated youngsters, workless and hopeless, venting their anger on the streets? Half the country, increasingly well-paid, spending their money to protect their wealth from the rising anger of the other half? Factories burnt down? The army called back from Ulster to maintain order in mainland cities?

Opportunity

Towns, cities and villages taking control of their own economies, reorganising local economic life and creating new jobs? A greater share-out of work, leisure and income? More education for all? Less emphasis on work as the centre of life, and more on the community? An end to dull, meaningless drudgery on the conveyor belt and a chance to make work rewarding and interesting? The beginning of a new way of life?

Well, it *could* happen. Before the last industrial revolution, who would have thought that the new factories could bring the affluence we know today? And the people who bring about this new revolution could be just the ones who today are valued least — the unemployed, with their free time, their wasted skills and their desperate need to get things changed.

> 'The stone that the builder refused
> Shall be the head cornerstone.' (Jimmy Cliff)

CRISIS TWO: PERSONAL

Danger

Long spells of depression? The feeling that everything is against you? Loss of energy and enthusiasm? Your marriage breaking up? Hopelessness? All your skills and energy going to waste?

Opportunity

A chance to stop and think? A chance to find out what you really want to do? A chance to do some of those things you always wanted to do? A chance to build a new secure basis for your life, which is not always so dependent on having to have a job? A chance to break out of old ruts?

Once again, it *is* possible. This Handbook tries to do two things:

- To give you the skills, information, etc. that you need to help you get a job.

- To suggest the ways in which the crisis can become an opportunity — for *you*.

I hope it does!

Chapter 2: An unemployment survival kit – building new supports

OUR BASIC SUPPORTS — REBUILDING THE SUPPORTS — Purpose and direction — Regular daily activity — Identity and self-respect — Companions and friends — Money — IN THE END: WHAT IT ALL COMES DOWN TO — SUMMARY CHECKLIST

OUR BASIC SUPPORTS

Why do many people feel depressed and 'down' when they are unemployed?

When you are in a job, or in regular training or education, your job gives you important personal support in five basic ways.

If you lose that job, or just don't have one, you may also lose these important supports. The result: Depression.

The basis of this short survival kit is therefore to:

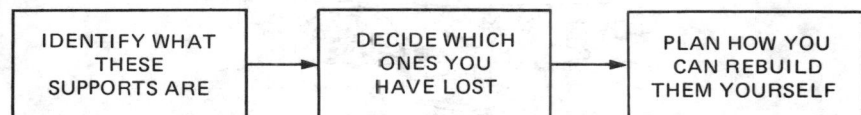

IDENTIFY WHAT THESE SUPPORTS ARE	→	DECIDE WHICH ONES YOU HAVE LOST	→	PLAN HOW YOU CAN REBUILD THEM YOURSELF

These are the five basic supports:

1. A job brings a sense of *purpose and direction*. You know what you are doing, and why you are doing it.

   When people lose their job, they often feel that they have lost their sense of purpose and direction. They may feel that they are drifting, unsure of where they are going.
   - Is this true for you? Yes ☐ No ☐ (Tick which is right for you.)

2. A job brings *regular daily activity*. It gives us regular work, regular hours, regular meal-breaks and regular holidays. Take the job away, and every day tends to become the same, week in, week out.
   - Is this true for you? Yes ☐ No ☐

3. A job provides a personal sense of *identity and self-respect*. You can say, 'I am a steel-worker', 'I am a nurse' or 'I am a student'. You know who you are, and you can tell other people who you are. It feels good. (Unless you hate the job.)

   Take away the job, and you take away this identity. 'I am unemployed' just doesn't sound so good, somehow, and most people hate saying it.
   - Is this true for you? Yes ☐ No ☐

4. A job provides us with *companions and friends*. This is so obvious that we

take it for granted. But take the job away, and you quickly realise how much you miss them. Without these friends, many unemployed people are alone for much more of the time, which can get depressing.

- Is this true for you? Yes ☐ No ☐

5. A job provides *money*. For most people, a job brings in more money than they get from Unemployment Benefit or Social Security Benefit.
   - Is this true for you? Yes ☐ No ☐

These supports are very important to us.

If we lose them, it is no wonder we feel down. *HELP!*

In Figure 1 below, ask yourself just how much support you feel you were getting from each of these five items in your last job (or school, college, etc.), and fill in your scores in Line A, using this scoring system from 5 to 0:

5 = a lot of support.	2 = not much support.
4 = quite a lot of support.	1 = hardly any support.
3 = satisfactory.	0 = no support at all.

Then do the same in Line B, by asking yourself how much support you are getting from each of them while unemployed.

	Purpose and direction	Regular daily activity	Identity and self-respect	Companions and friends	Money	Total
Line A: Last job etc						
Line B: Being unemployed						

Fig. 1. How much support?

Your figures can tell you two things:

1. By comparing your two totals in Lines A and B, you can find out how much being unemployed has thrown you off your balance by taking away or weakening your supports. (You might find that your overall sense of support has *increased* in unemployment.)
   - Points lost or gained in unemployment: . . . . lost/gained

2. You can see which particular supports need rebuilding. This is the basis of survival. You should pay serious attention to any support where you scored 0, 1 or 2 in Line B, and then plan how you can rebuild these supports (see below).
   - Supports need rebuilding: . . . .

        . . . .

        . . . .

        . . . .

        . . . .

REBUILDING THE SUPPORTS

1. A sense of purpose and direction

The key to rebuilding this support is to set your own goals and to make your own plans.

You may be lucky, and find a new job quite soon. As unemployment rises, however, the average time that it takes an unemployed person to find new work or training is increasing, and it is possible that you might be out of work for several months.

If you tell yourself, 'Oh, I'll have a job by next week' you will not be able to make any plans for activities that last more than a day or two. If you recognise that you might be out of work for several months, then you can say to yourself, 'I have got several months all to myself. What can I do with this time?' and then make your plans accordingly.

It is worth spending some time on this planning. You could perhaps take a sheet of paper (or a page in the Jobhunting Journal — see Chapter 4) and ask yourself, 'What are my three main priorities?' and then perhaps, 'What would I most like to do in this time?'

- Your main activity may be constant jobhunting. Chapter 4 has advice here, and will enable you to make thorough plans.
- Some people decide to use the time to take up a new skill, which they either learn on their own or at a class. See Chapters 6 and 7.
- There are numerous kinds of neighbourhood activity where there is always room for people wanting to help. Chapter 8 has suggestions and ideas.

- You are allowed to attend college for up to 21 hours a week without this affecting your entitlement to benefit. You could start studying for some exams or certificates. See Chapter 7.
- Some people use the time to start building the foundations of what later turns into a small business or self-employment. Chapter 9 has advice.

What are your own plans?

Set yourself overall targets for the things that you would like to do during the next four months. In Chapter 4 I suggest that unemployed people keep a Job-hunting Journal to help in the general task of finding a new job and rebuilding supports. When you have made your Journal, take a page and draw up a list of your plans, as in the table below.

March — July 1982: goals

- One hour per day routine jobhunting work.
- One evening per week helping at the youth centre.
- One afternoon per week working at the neighbourhood advice centre.
- Teach myself woodwork. (Is there a class? Who could help me? Could I do something in return if Ted helped me for one evening a week?)
- Go swimming twice a week.
- Spend some regular time reading.
- Join a rambling, hiking or mountaineering club and go away for a few weekends.

Draw up weekly timetables in advance
Arrange to spend an hour on your own, perhaps every Sunday evening, planning out the next week and checking off the things you did or didn't do in the last week. Use pages in your Journal to do this.

2. Regular daily activity

If you get into the habit of drawing up a regular weekly timetable, you may find that this is enough — or that it includes a daily timetable. By planning your day *in advance* the night before you will be able to get a lot more done during the day. Figure 2 gives a couple of sample daily timetables.

8.00 am — Get up, help get the kids off to school.
9.00 — Clean the kitchen floor.
9.30–11.30 — Work through the job advertisements in the daily papers, magazines, etc, and write off any application letters.
11.30 — Phone the Jobcentre, to check if anything has come in. (New jobs arrive every day, and have been sorted by mid-morning.) Until lunch — working on the allotment, or taking the dog for a good walk.
1.30 — Lunch.
2.00–3.30 — Helping out on the neighbourhood housing action project, on the rota.
3.30 — Collect kids from school, have tea.
4.30–6.00 — Helping with gym/football/netball coaching at the school.
6.30 — Cook the family supper. Time with the children.
***9.25 — Starsky and Hutch.

7.30 am — Get up, take cup of tea back to bed, read good book in bed, and plan the day.
8.30 — Get up, breakfast.
9.30 — Go to the library to check through the papers, making a note of any suitable jobs. Call in at the Jobcentre.
10.30 — Swimming pool with Kate.
11.30 — With Kate to visit Pete and Jenny. Lunch with them.
2.15 — Commerce class, at Tech.
4.00 — Home, tea, sit down to write off any application letters, etc. Make any phone calls to ask for further info on any advertised jobs.
7.30 — Meeting of local chess club.

Fig. 2. Two daily timetables.

3. Identity and self-respect

This may be the hardest support to rebuild. It is also the most important. Chapters 3 and then 8 might be helpful to you if you need to build up your own sense of identity and self-respect.

If you tend to see yourself mainly in terms of your job, so that you need to be able to say, 'I am a . . . .' you may find it hard if you cannot say that any more. (You might even find yourself still saying it, and not admitting to being out of work.)

If, on the other hand, you see yourself more (or also) in terms of your family, your friends, your activities, your regular place in the pub and your place in the local football team you will probably not be so bothered, and your self-respect will not be so affected by your being out of work.

How many ways are there in which you can happily say, 'I am . . . . '
(e.g. 'I am a father of two children — married to Jean — a keen gardener — a champion onion grower — a native of Neath — a Swansea Town supporter — a member of the youth club committee — a person who loves mountain-walking')?

'I am . . . .

. . . .

. . . .

. . . .

One of the main ideas in this Handbook is that it is important for people to build up their identity outside their jobs. If we find ways in which we can gain fulfilment and satisfaction by doing things locally in the community, and on our own, we will not be caught off-balance so much when we lose our jobs or have to spend six months in between jobs.

You may need to take on some new activities in order to build up your iden-tity and self-respect (use Chapter 8). Then, when someone asks you that old question, 'What do you do?' you can answer, perhaps, 'Well, I am out of a job at the moment, so I can't say much about that. I teach swimming to the scouts on Fridays, I'm learning welding at the Tech on Tuesdays, I'm quite involved in our local tenants committee, which is trying to get some changes made round here, and I am learning to play the bagpipes.'

By doing this kind of thing you are saying to yourself and to the world that just because you have been made unemployed does not mean that life is going to get you down. The hell it will.

4. Companions and friends

Companions and friends are really important. One of the really depressing things about being unemployed can be not meeting enough people and spending too much time on your own, or just with the family at home (which can also cause extra worries and tensions).

The more things that you do with your time, the more people you will meet. This can have an extra benefit because many jobs are found on the grapevine. The more people you get to know and see regularly, the greater your chance of hearing about something.

If you are a rather shy person who prefers to do things alone, such as reading or gardening, you might need to make a deliberate effort to take up some new kind of activity where you *will* meet people, even if it makes you uncomfortable at first.

Make a deliberate point of calling in to see your friends, just to keep in touch.

You could also join or form a special support group of people who are also out of work, so that you can help each other in various ways. Use Chapter 5 to help you.

5. Money

This is obviously the hardest support to rebuild. You may not be able to do much about this immediately but do check with Chapter 10 (about benefits) which may have one or two surprises. For instance, if you are getting Unemployment Benefit, are you also claiming a rent rebate or allowance? You should be. Have a look at Chapter 9 as well.

IN THE END . . . . WHAT IT ALL COMES DOWN TO

It is one thing to read a Handbook on how you can turn being unemployed into a constructive opportunity, and it is another thing to start *doing* something about it. In the end, it all comes down to one simple thing — do I do something, or don't I do something?

There is a time for wondering what to do next, a time for making plans and then there is a time for action. It is worth spending some time on the planning stage. There is no need to hurry. Give yourself time to think about things properly so that when you decide to act you are happy that you are taking the right step. After all, it could be quite a big step, when you look back on it later.

SUMMARY CHECKLIST

Make a list of the supports that you need to rebuild.

Then take a sheet of paper for each support, and write a list on each sheet of the things that you plan to do to rebuild that support. After a month you could do the exercise at the beginning of the chapter again, and see:

- Whether your scores under each support have improved.
- Whether your total score for how you are surviving while unemployed has improved.

Even a very small improvement will be something worthwhile. In this way it is possible to build up your own life-supports so that you are no longer so dependent on having a job for your personal security. Your life should feel much more stable and satisfying.

Chapter 3: But what shall I do?
Discovering your natural skills

INTRODUCTION – WHAT ARE YOUR NATURAL SKILLS? – RELATING YOUR SKILLS TO POSSIBLE JOBS – GETTING DIRECT EXPERIENCE

INTRODUCTION

- Do you know what kind of job you would like to do? Yes ☐   No ☐
- Do you know what kind of training might help? Yes ☐   No ☐
- Do you know what you are good at doing? Yes ☐   No ☐

If you can answer 'Yes' to all these questions, you might want to skip this chapter. However, it might tell you something about yourself and your abilities that you did not realise.

To work through the chapter properly, find a quiet place where you will not be disturbed. The chapter can be worked through alone, by two people working together or by a small group. If you can find a friend who is also interested you might benefit by doing it together.

Equipment: an empty notebook, at least 5 in.x 8 in. in size (the size of this Handbook). This can later become your Jobhunting Journal. If it has strong covers, so much the better.

What do you look for in a job?

It will help you if you are clear about this. There are seven items listed below. Which matter most to you? Score them in order of importance, so that the one that matters most scores 7, and the one that matters least, 1.
- Pay. . . . .
- Fulfilment and satisfaction . . . .
- Pleasant company and conditions. . . . .
- Opportunities for advancement . . . .
- Responsibility, initiative, independence . . . .
- A chance to exercise skills. . . . .
- Security. . . . .

If you think that any are equally important, give them an equal score.

In your wildest dreams. . .

Before starting, take five minutes to ask yourself if you have already got some private ideas about what you would *really* like to do. They may be ideas about which

In your wildest dreams. . . . .

you feel, 'Oh, but that is ridiculous' — well, that doesn't matter. Try to think of four or five jobs that you would love to do if you had all the skills, money or whatever else was needed. Write them in a list at the back of your Journal, and then forget about them for the time being.

Your natural skills

Inside everyone there are special talents and abilities. This chapter should help you to find out what your own are, and then find ways in which you can develop them, either by finding a job (hopefully) or activities outside of a job which will allow you to express them. You may have:

- A special skill with wood, or a knack with electronics.
- A natural ability at organising.
- An inborn 'way' with plants, or animals.
- A natural eye for beauty, colour and design.
- A natural ease with other people, and an instinct to care.

In an ideal world, everyone ought to be able to do work that expresses his or her natural skills. In *this* world (until we make it better), most people have to put up with less. We still have the abilities, however, and we owe it to ourselves to find out what they are and to *use* them.

WHAT ARE YOUR NATURAL SKILLS?

(The exercises used here are adapted from the book *Job Power Now!*, by Bernard

Haldane, a career counsellor in the USA, who is acknowledged with thanks. (Acropolis Books, Washington DC 20009 USA.)

The steps that follow help you to identify what your own natural skills and abilities are. They do this by reference to the things that you have done, in your life, that gave you a special sense of satisfaction or achievement.

Step 1

On Page 1 of your notebook make a brief year-by-year diary of your life so far. Write down the important things that happened to you, the important things that you have done, the jobs you have held and the schools you went to. The purpose of this is simply to re-awaken your memory. (Approx. 15 minutes.)

Step 2

Go through your life again, more slowly, year by year. On the next few pages of your notebook write down *the things that you have done that you feel proud of, the things that gave you a special sense of satisfaction and the things that gave you (or give you) a specially deep sense of enjoyment.* Take it slowly. You will begin to remember things that you had completely forgotten about. Carry on until you reach the present. You might end up with a list of 10-15 items, depending on how able you are to admit that you were proud to have done something, or that some experience meant a great deal to you. Many people find this hard, because they find it difficult to believe that they have done *anything* to be proud of — which is never true. (Approx. 30-60 minutes.)

If you think of something, but then think, 'Oh, but that doesn't count', don't squash the idea. Write it down. Don't only measure your experiences by the things that school or society normally counts as an 'achievement'. It is what *you* feel to have been an achievement or an important experience that matters, not what others think.

Typical achievements that people might feel especially proud of:
- Laid a crazy paving terrace on my own.
- Made a radio from a kit.
- Ran a successful sales campaign.
- Made a pot.
- Completed a five-day hike with friends.
- Organised help and looked after people when there was a car crash outside my home.
- Supported a friend over a month when she was feeling terribly low.
- Designed the stage set for a pantomime.
- Wrote a special paper on a subject that fascinated me.
- Worked successfully as foreman for a year.
- Learnt how to handle and to master a new machine.

Step 3

Look through your list when it is complete and choose the seven achievements

Skills	1	2	3	4	5	6	7	TOTAL
Skills with animals, or plants Analysis Artistic skills								
Caring Communication Co-ordination								
Creative Constant hard work Dealing with people								
Design skills Diplomatic Efficiency								
Eye for detail Figures Food skills								
Healing Human relationships Ideas								
Being imaginative Initiative Being inventive								
Leadership Listening Management								
Good memory Money skills Organising abilities								
Outdoors Patience Performance								
Persistence Planning Practical								
Promotion Reliable Repair skills								
Research Science Self-discipline								
Selling Solving problems Stamina								
Talking Teamwork Thorough								
Tolerance Understanding machines Words								
Work with your hands Writing								

Fig. 3. Skills and achievements.

which you feel mean the most to you. Try to place them in order, numbered 1 to 7.

Step 4

Using the list of skills in Figure 3, take each of your seven achievements in turn, and tick off the skills which you think you were using. Give your first three achievements double ticks. This is a very good exercise to do with someone else, since you could help each other.

Step 5

When you have finished, add up the number of times you used your skills in the TOTAL column (counting two ticks each for the first three achievements). Then take the *five* skills with the highest scores, and copy them down into your notebook under the heading 'Natural Skills', with their scores next to them.

RELATING YOUR SKILLS TO POSSIBLE JOBS

Take a break. If you can, go for a walk to let your mind have a rest. Let it quietly mull over what you have just done.

The problem now is this: which kind of job would give you a chance to develop your natural skills? There are thousands of different kinds of job, some of which you may never even have heard of, let alone know what they involve. By the end of this chapter you should end up with a list of about five possible jobs which you can then explore in more detail.

From now on, you need to be asking yourself constantly over several weeks, 'What kind of job?'. If you search constantly and persistently enough, your answer will come.

It might be that you are not able to find a job which gives expression to your skills at all. Not everyone is able to. This need not matter. It *does* matter, however, that you find some way to develop and express your skills. There may be things that you can do in your own time, or on a voluntary basis, which will give you the fulfilment which your own special abilities can offer you. Use Chapter 8 to help you decide what you want to do. The point is — just because you are unemployed need not mean that you stop doing anything. You can still press ahead, and find your own fulfilments from life.

Step 6

On a fresh page in your notebook, start to keep a list of 'Possible Jobs'. Whenever you think of one that seems interesting or possible, write it down straight away. You can investigate it later.

Suggestions to help you build up your possible jobs list
1. Make an appointment to see someone at the Careers Office.

Show them your list of skills, and ask for their suggestions about possible jobs. While you are at the Careers Office, give yourself time to look for other ideas in the magazines there.

2. Have a look through the book *Equal Opportunities: A Careers Guide for Women and Men*, by Ruth Miller (Penguin £1.95 — in libraries).

3. Talk to your friends about your search:
   - The friends who are eager to help you with ideas and suggestions.
   - The friends who are good at listening. If you can find just one person who will listen to you while you talk about what you might like to do, you may find that you discover new ideas yourself.

4. If you are in a group (see Chapter 5), or have friends in a similar situation, get together to help each other with ideas.

5. Scan the papers regularly, and ask yourself if the jobs being advertised might fit your skills. Then phone up to find out more about them.

6. Try to ignore any voice inside that says:
   - 'But no one is ever paid to do that.' (You never know.)
   - 'Oh, I could never do that, it needs a certificate in fire-eating and competent Arabic.' (You could learn fire-eating from a travelling theatre-group, and Arabic by a correspondence course, see Chapter 7.)
   - 'Oh, but women don't do that kind of work.' (Oh yes they do.)
   - 'Oh, but men don't do that kind of work.' (Oh yes they do.)

7. What about the possibilities of self-employment? Chapter 9 might give you some ideas here. There is always jobsharing, too — see Chapter 4.

Step 7

For this step, you need to have a list of *five possible jobs*.

Take five sheets of paper, one for each job (or five pages in your notebook). Then take each job in turn, and ask yourself these questions, writing down your answers:

1. Imagine yourself doing this job. What does it feel like?
2. Imagine yourself doing this job for a whole year. What does it feel like after a year?
3. If there is a change in how it feels, why might this be?
4. Which of your personal needs would this work be fulfilling?
5. Which of your personal needs would this work *not* be fulfilling?
   Do you have any negative thoughts about it?
6. Are there any changes which you could make to the work, or to the way you did it, which would meet the changes noted in Question 3 or your comments in Questions 5? (Do it part-time? Do it on a jobsharing basis? Wait a year or so? Don't do it as a job, but just as an activity? Ask a friend to join you, so that two of you could do it together? Etc.)
7. Is it worth following this idea any further?   Yes ☐   No ☐

If 'Yes', which steps will you take from those listed below?
- Visit the Careers Office to ask for advice and information.
- Read books, magazines about it. (Library, careers library.)
- Talk to someone who is already doing it. Who?
- Visit the job, to see just what it is like.
- Try to arrange a three-day trial, or find a way of actually trying it out. (See below.)
- Develop the work involved in the job on your own, with friends or at evening class, etc.

Coming up bump against reality
You will probably discover that some of your ideas are impossible for one reason or another. If this is really the case, ask the Careers Office for suggestions about jobs that are similar which you might not have thought of. Then add them to your list, and do Step 7 on them. It may also be that although you cannot do anything about an idea *now*, you may be able to in a year or two.

Coming up bump against reality.

We all miss real opportunities simply by not being bold enough. Stretch the limits of what you see as possible. The unexplored life is hardly worth living.

GETTING DIRECT EXPERIENCE

The best way to find out whether you like a job or not is to find a way of getting direct experience of it. There are several methods which you could try:
- *Getting a chance to work at the job for a day (or longer) without pay:* use

the Careers Service, the Yellow Pages and friends and contacts to help you. Find employers whose work interests you, then phone up or write to them to ask if you could come and find out something about the work. (See telephone advice, Chapter 4.) If you can get an introduction, it will make things easier. Make it clear to the employer why you want to visit, in case she thinks that you are trying to squeeze a job out of them. (You might find that they are impressed with your way of going about things, and *do* ask if you would like to work with them — but that is not why you are going. Anyway, do not let yourself be rushed into a decision.)

- *Local daytime and evening classes, local voluntary work, getting experience in your own time:* there are many ways in which people manage to press ahead with whatever it is they want to do, without waiting until they get a proper job. Use Chapters 7 and 8 for ideas.

- *The Manpower Services Commission:* the MSC runs 3 schemes which offer unemployed people a chance to get direct experience of work:

    The Work Experience Programme for young people under 19 allows you to work for up to 6 months at a job that you want to learn more about, with a weekly allowance. See Chapter 6.

    Wider Opportunities Courses are held at Skillcentres around the country. These offer people of all ages a chance to try their hand at a wide range of different tasks. See Chapter 7.

    Wider Opportunities for Women courses. See Chapter 6.

Step 8

What happens if you have several different jobs to choose from, and you cannot decide which one to concentrate on? This short exercise may help you to decide — it is also quite fun to do.

Items in order of importance	Q	Previous job		Possible jobs									
				1		2		3		4		5	
		a	b	a	b	a	b	a	b	a	b	a	b
. . . .	7												
. . . .	6												
. . . .	5												
. . . .	4												
. . . .	3												
. . . .	2												
. . . .	1												
Job total score		. . . .		. . . .		. . . .		. . . .		. . . .		. . . .	

Fig. 4. Scoring possible jobs.

1. Take the list of seven items at the beginning of this chapter, and copy them into the seven spaces in Figure 4, placing the one that mattered most to you at the top, etc.

2. Think of a job that you have done in the past. Go through each of the seven items in turn and give each item a score in Column (a) (previous job) on this basis: 4 = very good, 3 = good, 2 = satisfactory, 1 = not very good, 0 = bad.

3. Multiply each score by the figure opposite it in the Q column, and place the result in Column (b).

4. Add up the figures in Column (b) and this will give you a special personal score for that job.

Now repeat this process for each of the jobs on your 'possible jobs' list. When you have finished, you will have a *rough* indication which job might suit you best. But remember that there are always unexpected hidden factors about every job, such as the people you would be working with. The highest score that a job can get is 112; the lowest score is 0.

You can't find a job? All is not lost. Other ways of developing your skills.
If there really seems to be no way in which you can express your natural skills through a paid job (or at least, not for the moment anyway), work out ways in which you can express them through other activities. Use Chapter 8 to help you. Just press ahead with what brings you satisfaction, and you might find that in a few years time it leads to something; perhaps a paid job, perhaps self-employment. The point is — just because you can't find a job, it doesn't mean that your life has to stop. Press on regardless!

Chapter 4: Practical jobhunting made easy

INTRODUCTION — PREPARATIONS — Personal Information Chart — Curriculum Vitae — Jobhunting Journal — FINDING OUT ABOUT JOBS — Careers Service — Jobcentre — Employment agencies — Newspapers, magazines and journals — Writing letters 'on spec' — Visiting — The grapevine — Local noticeboards — Local radio — Specialist information centres — Jobsharing — APPLICATION LETTERS AND FORMS — THE INTERVIEW — TELEPHONE TECHNIQUE — OBSTACLES TO JOBHUNTING — Racial discrimination — Sex discrimination — Criminal record — Transport

It is much easier to put enthusiasm into hunting for a job if you know what you are hunting for:

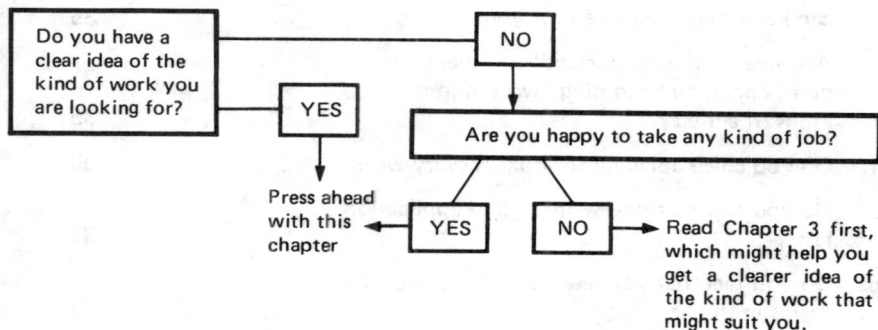

```
Do you have a clear idea of the kind of work you are looking for? ──── NO
                            │ YES
Are you happy to take any kind of job?
Press ahead with this chapter ←── YES    NO ──► Read Chapter 3 first, which might help you get a clearer idea of the kind of work that might suit you.
```

DO YOU NEED THIS CHAPTER?

Have a quick check through this exercise:

		Tick if 'Yes'	See page
1.	Have you written yourself a personal information chart or curriculum vitae?	☐	23
2.	Are you pleased with it?	☐	25
3.	Have you had copies made of it?	☐	25
4.	Do you have a notebook or a file where you can make lists and plans to keep your jobhunting organised?	☐	27
5.	Are you calling at the Careers Office or Jobcentre regularly, or phoning if you cannot call?	☐	28
6.	Have you registered with employment agencies? How many?	☐	29

		Tick if 'Yes'	See page
7.	Do you keep a daily check on jobs in the daily papers?	☐	30
8.	Do you keep a weekly check on jobs in the weekly papers?	☐	30
9.	Do you keep a check on jobs in any relevant magazines, journals?	☐	31
10.	Are you writing letters 'on spec' to possible employers?	☐	32
11.	Are you making visits 'on spec' to possible employers on an organised basis?	☐	33
12.	Have you told your friends and neighbours that you are looking for work, so that they can keep their eyes open for you?	☐	35
13.	Are you making new friends and meeting new people, by becoming involved in new kinds of activity?	☐	35
14.	Do you check local notice-boards every week?	☐	36
15.	Do you feel that you write a good application letter?	☐	37
16.	Do you feel that you always fill in application forms in the best way?	☐	41
17.	Do you feel confident about your interview manner?	☐	42
18.	Do you prepare carefully for an interview, by finding out about the job, thinking of questions to ask, etc?	☐	43
19.	Have you ever had any interview practice?	☐	44
20.	Do you feel confident about your manner on the telephone?	☐	46
21.	Do you have a small group of other unemployed friends, whom you meet regularly with, in order to help each other with your unemployment problems?	☐	49

TOTAL:

Scoring:

- *15 – 21:* you are doing well in your jobhunting efforts. It may be useful to check on the points which you did not score on, just in case there may be something useful to you there.
- *7 – 14:* you are doing quite well, and possibly as well as anyone else – but there is more that you could be doing to help yourself. Read through the chapter, or at least the sections covering the points on which you did not score.
- *0 – 6:* you have good reason to feel hopeful, because there is room for a lot of improvement in your jobhunting which might bring you success. If you read through the whole chapter and develop a systematic approach to looking for work your chances of getting a job will certainly improve.

INTRODUCTION

There are four stages of jobhunting to think about, each of which needs to be given proper attention:

1. Preparations

- Preparing a 'personal information chart' or 'curriculum vitae' which you can use with your application letters, when approaching employers directly and at interviews.
- *Keeping organised.* The Handbook suggests that you make yourself a Jobhunting Journal, where you can keep a careful track on the letters you send off, employers you have contacted, people to see, plans, lists, etc.

2. Finding out about jobs

There are ten different sources listed below, with advice and information on each.

3. The application letter

For every 100 letters that arrive after a job is advertised, 90 are thrown away the morning they arrive. Your own letter must survive.

4. The interview

It is easy to waste a good opportunity by not preparing carefully enough.

STAGE 1: PREPARATIONS

The personal information chart

It is very important that an employer has a chance to find out about your good points, your skills and your experience. An application letter has to be short, and application forms never allow you to say very much about yourself.

If you prepare a special sheet of information about yourself to give to employers when you are applying for a job, writing letters or visiting 'on spec', and at interviews, you will give yourself a much better chance of getting the job. The employer will know everything about you that you want him or her to know.

A 'Personal Information Chart' ('PIC') allows you to write down details about yourself, covering your experience, your work background, your skills, your abilities, your qualifications, and anything else you want to say about yourself. Chapter 3 can help you to know what your own natural skills and talents are, with evidence from the things you have done which demonstrate these skills.

Preparing a PIC

Take some scrap paper and give yourself at least two hours.

If you can arrange to do this with someone else who is also unemployed, so that you can help each other, so much the better. This is also something that a small group can very usefully do together. (See Chapter 5.)

There are six headings which you can use on your PIC. Write it in rough first, before trying to write it out perfectly.

PERSONAL INFORMATION CHART

1. *Your name, address and telephone number*
   If you aren't on the telephone, do try to find a phone that you can give as a *contact number*. Ask friends and neighbours. (If they can't help, the local Social Services Department might put you in touch with someone who is housebound, who could take messages, and who would welcome regular visits.)

2. *'Experience', 'work experience' or 'employment history'*
   - Choose the heading you prefer.
   - Make a list of *all* the jobs (paid or voluntary) and activities that you have done.
   - Choose the best 5 or 6 and write them in your PIC. (You can leave out anything you didn't enjoy, etc.)

3. *Skills and abilities*
   - Check Chapter 3, and write in skills from it.
   - Write in any practical skills (welding, typing, painting, child care, etc).
   - Give examples of when you've used each skill.
   - List all the equipment you can use (calculator, switchboard, sewing machine, etc. — at home, in a factory, in the garden, etc).
   - Choose the best of the skills, and write them in your PIC.

4. *Personal achievements*
   - Describe anything you have done that you feel particularly proud of. (Use Chapter 3 to help.)

5. *Education*
   - Give details of your schools, college, exams, qualifications, etc.
   - Mention any other education — evening classes, courses, projects, etc.

6 *Referees*
   - Give the names and addresses of two people who know you well and will write well of you — but not from your family.

Fig. 5. Personal Information Chart.

Leave it for a day or two, in case you think of anything else, and get a friend to check it too before you write it out. Make it as tidy and clean as you possibly can. Get it typed, if you like.

Making copies of your PIC
It usually costs between 6p and 10p a copy to use a photocopying machine commercially. You really need about 100 copies of your PIC, which might prove a bit expensive. Here are some suggestions:
- Print 10 at a time, and print more when you have used those.
- Ask a friendly estate agent, printer, building society, or a friend who works in an office, if they could do it for you at cost price.
- Ask the Careers Office, Jobcentre or youth centre if they can print copies for you cheaply.

Making copies of your PIC

How should you use your PIC?
- Include a copy with application letters for jobs.
- Take copies when you go 'visiting' possible employers. (See below.)
- Take a copy to your interview.
- Ask your friends if they would take a few copies to give to anyone they know who might have a job going, or know about jobs going. (See below, 'The grapevine'.)

The curriculum vitae

This is for professional and managerial jobs. (Also sometimes known as 'Career History'.) The procedure is more formal. First work out what you want to write in rough, and show it to a friend for comments, before typing out the final version. It is important to make a first-class job of your c.v., since the impression it gives will be an impression of you.
    The headings you should work with are:
- Name, address and telephone number.
- Your qualifications.
- A year-by-year run-through of the positions you have held, the work they involved, and the responsibilities which you held.
- Other information — but avoid bringing in too many details of hobbies, etc. which are not relevant to applications for professional and managerial jobs.

```
Job Title.......................................... Ref. No................
Consultant........................... Location.............................
```

CAREER HISTORY

PERSONAL DETAILS

Name in full; John Joseph SMITH

Present address; 100, Main Close, Upton, Huntshire RO5 LMS

Telephone No;(Home) Upton (3305) 777, (Day time) Selford (7053) 9922

Date of Birth; 7th April, 1940. Marital Status; Married, 2 children 11 and 9 yrs

EDUCATION AND QUALIFICATIONS:

1952 - 58	King Edward VI School, Portchester GCE 7 'O' levels: 4 'A' levels French (Grade B) Physics (Grade B) Chemistry (Grade B) Botany (Grade C)
1958 - 61	Midlands University BSc 2 (i) Chemistry
Professional:	Member of the Institute of Sales Management

EXPERIENCE AND ACHIEVEMENTS

May 1974 - to date	U.K. Paint Co. Ltd., White Lane, Selford (Specialist Paint and Varnish Manufacture) General Sales Manager responsible to the managing director for sales of paints and varnishes to the building, automotive and furniture industries, in the U.K. and parts of Europe. Staff of 75 representatives and 40 office staff. Also responsible for recruitment, training, sales and expense budgeting. Built up sales from £40M to over £85M in the past 5 years, set up and trained special teams to cover EEC markets.
June 69 - March 74	Jordan and Brush (Paints) Ltd. Cover Way, Bramley. (Chemical and Paint Distributors). Sales Manager (paints) responsible to the general sales manager for sales in the UK to specialist retail outlets. Staff of 15 representatives and 8 office staff. Built up sales of over 20% a year; negotiated several contracts in excess of £½m and was responsible for setting up the company's new pricing structure.
Sept 63 - June 69	Barmeston-Meeds UK Ltd, Thornton (Synthetic Resins and Plastics) Technician/Assistant Laboratory Manager responsible to laboratory manager for quality of all incoming raw materials. Staff of 3.
62 - 63	Brampton Private School - Teacher (Chemistry, Physics and French)
OTHER INFORMATION	1974:8 week management course - Park Gate Management College Fluent French - both written and spoken; attend EEC conferences. I have written technical articles for journals and conference papers.

From *Executive Post.*

Fig. 6. An example Career History

Only mention information which demonstrates your skills, abilities and responsibilities.

Use white A4 paper, clearly typed, and well spaced on the page. Figure 6 gives a sample c.v.

Jobhunting Journal

One way to organise your jobhunting efforts, is to make yourself a Journal — a notebook, at least 8 inches by 5 in which you can keep a track on various different things to do with your jobhunting and your time while you are un-employed.

Your Journal might include any or all of the following:
* The exercises suggested in Chapter 3 to help you discover your own natural skills, and possible jobs which might suit them.

Jobhunting Journal

* Your week-by-week plans. (See Chapter 2.)
* Notes for your PIC or c.v.
* Addresses and phone numbers (e.g. Jobcentre, Unemployment Benefit Office, Careers Office, DHSS, etc.).
* Checklists of your regular jobhunting efforts.
* Lists of employers to write to or visit.
* Names of employment agencies to keep in touch with.
* Lists of possible 'grapevine' contacts.
* Lists of advertisements answered and application letters posted.
* Notes about application letters, and a copy of a good 'standard' letter.
* Notes about firms, for interviews.
* Lists of your plans and priorities over the next few months or weeks. (See Chapter 2.)
* Lists of 'possible choices', with reasons for and against.

Number each page, and make an index of contents at the front. This will help you to feel that your Journal is valuable and useful.

Reminder: Personal decision to keep a Jobhunting Journal: Yes □ No □.

STAGE 2: FINDING OUT ABOUT JOBS

There are many different methods of finding out about jobs. Most people ought to be able to use several of them. However high the level of unemployment, there are always new jobs coming up, simply because people constantly change their jobs, just as they change their houses. Only 23% of people looking for work find their jobs through the Employment Office or Jobcentre. The rest all find work through *other ways*, and if you ignore these other ways, you will have a much smaller chance of finding a job.

Here are the details of the various ways:

1. The Careers Service

The Careers Service is run by the local county or city council, and is completely independent of the Jobcentre.

The Careers Service works in two ways that are relevant here.

- It can offer help and advice to young people who have left school within the last two years, and try to arrange interviews for jobs for them, when it has jobs available. There is detailed advice for young people on how to get the best out of the Careers Service in Chapter 6.
- It can offer help and advice to *anyone* who is unemployed and who wants to talk over ideas about possible jobs with someone who knows what kind of work he or she is interested in. They cannot arrange any interviews for adults, but their ability to offer advice should not be forgotten.

Don't just call in on spec, unless you want to browse through their literature and magazines. Phone first to make an appointment.

2. The Employment Office, or Jobcentre

(In case you are muddled, these are both the same thing, under different names. They are run by the Employment Services Division of the Manpower Services Commission. Employment Offices are gradually being changed into Jobcentres as they get their new carpets installed. The new Jobcentres are always in a High Street or shopping precinct. I use the term 'Jobcentre' for both here, and throughout the Handbook.)

If you go to the Careers Service, you will get personal service. At the Jobcentre, you will not get personal service unless you actually ask for it. The Jobcentres work on a 'self-service' basis which many people in fact prefer.

The basic details of all available jobs are typed onto cards which are displayed around the walls. If you see a job which interests you, ask about it at the counter, and you will be given the full details. You may be given an appointment then and there to go for an interview. There will be no need to write an application letter, but you may have to fill in an application form when you get there.

When you first register, you will be asked what your occupation is. You will then be put down in the files as looking for this kind of work, and if something comes up, you will be told about it. This sounds good, but it does also mean that if something *outside* your stated area of work comes up, you will not be told, so think carefully about how you want to be listed. You might want to put yourself on file as looking for two or three different types of work. Self-registration forms are in use in some offices. You fill one of these in yourself before the registration interview.

You will be interviewed by one of the 'Employment Advisers' when you first sign on, and then every so often, later. Your interview should last about 20 minutes. You may only get a quick five minutes, with a few brief questions about the kind of work you want. If you feel that you have been treated a bit cheaply, you can complain, and ask for proper treatment, but it may not help.

The Government spending cuts have also affected Jobcentres, and they are being forced to cut back on proper interviews.

The staff at the Jobcentre are also well informed about:

- The training schemes available with 'TOPS'. (See Chapter 7.)
- Jobs under the Government's special 'Community Enterprise Programme' (CEP) for unemployed adults. These jobs are advertised in the normal way, along with other jobs, which is why there is no special section on CEP in the Handbook.
- Special schemes for the disabled. (See Chapter 6.)
- Special schemes for people looking for jobs away from home. (See Chapter 11.)
- Special schemes for people with certain skills that are in demand, chiefly engineering skills.

Early visiting: details of new jobs arrive at the Jobcentre each morning, and are usually put up on the boards by 10.30 or so — so this is a good time to visit.

You could ask the Jobcentre staff if they know about any local work that is available, but not advertised, such as local labouring.

What if you live some distance away from the Jobcentre, and cannot call in very often? You can still phone up regularly. Ask when is the best time to do so. When you get to know them, you might be able to persuade them to phone *you* up if any suitable jobs appear. If you have to phone from a call-box, ask them to call you back.

Sometimes Jobcentres put details of jobs up in libraries, to help people who live some distance away. Always phone the Jobcentre about these jobs *before* getting in touch with the employer directly, since the cards are sometimes out of date.

- The staff at Jobcentres have learnt from experience that it is not worth their while to push anyone into a job against his or her will. The 'Unemployment Review Officers' appointed by the Government to 'encourage' people to get off the dole are attached to the DHSS and not to the Jobcentre.
- It is up to you to keep in constant touch with the Jobcentre. If you assume that they will look after you, nothing will happen.

3. Employment agencies

It is difficult to generalise about the private employment agencies, since they vary a lot. Some are very good, and treat people very well. Others unfortunately have a reputation for being a bit of a rip-off. The only way to find out is to visit them yourself, and ask your friends about them. If you do not use the agencies, you may be reducing your chances of finding work. Many people find that the way the agencies operate suits them excellently, offering them different kinds of work with regular variety. The agencies want to develop lists of good, reliable workers, and the people who are satisfied with the agencies are also the ones who are regularly kept in work.

If you look up 'Employment Agencies' in the Yellow Pages, you will find the local and regional agencies listed. You will see that some specialise in secretarial or driving work, etc. while others are all-purpose. The more agencies you are registered with the better, so ask every relevant agency to put you on their books. You will normally be sent a form to fill in, asking the kind of work you are willing to do.

Some agencies (e.g. 'Manpower') employ their staff directly and subcontract their workers to employers, while others live by the fees that the employers pay.

As a word of warning — you should be clear what the arrangements are at each agency about pay, sickness, dismissal, holidays, etc. Many people who live by work from agencies are not trade union members, so you cannot always depend on a union to get you out of a fix.

Private employment consultants
There are a number of private firms which exist by helping people to find work (mostly professional and managerial people). There is a section in Chapter 6 which covers jobhunting advice for this group of people. Certain groups of people can also get help from specialist organisations such as the Armed Forces Resettlement Information Office and the Institute of Directors. The British Institute of Management runs its own advice and information service, and a counselling service for members who may be facing redundancy or a change in their career.

4. Advertisements in newspapers, magazines and journals

This is perhaps the chief method by which most employers advertise their vacancies. You will probably need to cover several local papers every week, if you are to see all the jobs coming up. If this is going to cost a lot, you may do better to make a regular trip to the local library, where they will have the papers.

- Names of newspapers to check daily: . . . .

. . . .

. . . .

. . . .

- Names of newspapers to check weekly: . . . .

. . . .

. . . .

If you think a job might go very quickly, reply at once, but you can afford to wait for a day or two for most jobs, so that you have time to make sure that your letter of application is just as you want it. Most employers wait for all the applications to come in before deciding who to call to an interview.

Phoning for more information
Newspaper advertisements are always very brief, because of the cost. Before sending off your letter of application, check that you have enough accurate information about the job. Use the list below to check which things you know about, and then

phone up to ask about the things you don't know. Copy the list into your journal, so that you can use it regularly.

- Where is the job?. . . . .
- What type of work is it? . . . .
- Are there any age restrictions? . . . .
- Are any qualifications needed?. . . . .
- Would training be given?. . . . .
- What kind of a firm is it? . . . .
- What are the hours and the pay?. . . . .
- How do you apply? . . . .

Choices . . . .

You may find yourself speaking direct to the employer when you phone, so be very polite and friendly, and be careful not to say anything which might write off your chances. The job might, for instance, require a special skill at riding camels. Don't say that you can't straight away — rush down to the local zoo quickly, and take lessons. When it comes to the interview, and the employer gets to meet you face to face, he or she may decide that it doesn't matter that you aren't very good at it yet, whereas if you had admitted to it, you might have ruined your chances.

Magazines and journals
As well as the daily and weekly newspapers, there are magazines and journals which carry advertisements. Some are general, such as *Jobs Weekly*. Some are local, some are specialist, such as:

- *The Lady* and *Nursery World*, for jobs with children.

- *New Society, Community Care* and the *Health and Social Service Journal*, for jobs in social service and youth work.
- *Accountants Weekly, Recruitment Accountant, Marketing Week, Marketing*.
- Trade journals, for jobs in catering, grocery, hairdressing, etc.
- *The Times* and the *Guardian*, for occasional 'one-off' jobs, like working as a masseuse in the harem of the Sheik of Arabi, or as a crew member on a Bahamas-cruising yacht.
- *Exchange and Mart* carries advertisements for books which offer to tell you everything about work on the North Sea oil rigs, at sea, or overseas.

Remember: you can place advertisements in these papers yourself.

News items

It is worth keeping an eye open for news items which might tell you about possible jobs, such as 'Camels-for-children firm moving to town' which might have a job for you with your newly acquired skill in camel riding, or '£½ million granted for conservation work' which would definitely imply that someone is going to be offering work soon.

5. Writing letters 'on spec' (The direct approach)

You can write directly to employers, on the off chance that they might have a vacancy coming up. If they haven't, you can ask that they keep you in mind, if something does come up.

How can you find out the names of possible employers?
- Ask the Careers Service.
- Use the Yellow Pages.
- *Jane's Major Companies of Europe* (in libraries. You may have to order it, in small libraries).

Using a page in your Jobhunting Journal, make a list of all the firms which you plan to write to, and make a column where you can tick off when you wrote to them (noting the date), and note what kind of response you receive.

Your letter

Your letter will need to be interesting. It will need to make an employer think that you are genuinely interested in working for his or her firm, and that it is *not* simply one of 50 identical letters that you have sent off to firms all over the place. Well, it probably will be, so what can you do about this?

Your letter *must* be individually typed or hand-written. There are no two ways about this. If you think, 'How much quicker it would be just to write one letter, and then get it photocopied' you are simply wasting the cost of postage and your own time. They will end up in the wastepaper bin.

The key to all the methods of application for a job is to put yourself in the employer's position, and imagine what 'you' would think, if you received your letter, or form.

The photocopied letter comes as an insult, and makes the employer think, 'This person can't be bothered to write me a personal letter, so I can't be bothered to read it.'

If you have written a PIC or c.v. (see above), you can write a short personal letter, and include a copy of the PIC/c.v., which no employer will expect to be individually written out.

Figure 7 is a sample of a possible 'direct approach' or 'spec' letter.

'Address
Phone No.

November 10th 1981

Dear Sir or Madam,

I am writing to you tentatively, to enquire whether you might have any vacancies.

I have worked with Extoll Ltd on the assembly line and in the cutting shop for the past 5 years, until they dismissed 300 workers this spring. I was satisfied with the work there, and would be pleased to take similar work with your firm, if there were any available.

I am including a 'Personal Information Chart' with this letter, which gives a wider picture of my abilities and experience.

If you do have any vacancies, either now or in the future, I would be grateful if you could bear me in mind when selecting new employees.

Yours faithfully,

(Signature)

(NAME IN CAPITALS)

Fig. 7. A 'spec' letter.

6. Visiting

There are some employers who hardly ever advertise their vacancies, because they prefer to wait until people come to ask if there are any vacancies, believing this kind of person makes a more enthusiastic and reliable employee. Obviously, when using this method you may draw a lot of blanks; you'll be lucky to find an employer who has a job there and then. So *take copies of your PIC* with you; you can leave one with the employer and when a vacancy *does* come up she may contact you rather than advertising (which can be expensive).

If you make your visits with this idea in mind, then you need not feel depressed and gloomy if your hard day's work seems to bring you no joy. You have been out sowing seeds, and one of them might bear fruit later. It's like an investment in your own future.

Making plans

You can find out the addresses of possible employers by asking at the Jobcentre,

Making plans — take something to read.

at the Careers Office, and by using the Yellow Pages. Then sit down with a local map (and a bus timetable?) and plan your day carefully. There are these points that you should remember:

- Dress as you would for an interview — smart (but not *too* smart).
- Take something to read, in case you have to wait around a lot.
- Take enough copies of your PIC so that you can leave one with each employer you visit. Keep them in a folder, so that they don't get dog-eared and scruffy.
- Take a sheet with your other personal details on it, in case you are asked to fill in an application form — your NI number, details of your previous employment (with addresses and dates), prepared reasons why you left the jobs (see below, 'Application Forms') and your qualifications, which you might not have put in your PIC.
- Ask to see the right person. The receptionist will tell you who is responsible for recruitment, and then you should ask if you can speak to him or her for a few minutes. They are sure to ask, 'Have you got an appointment?' Don't just answer, 'No' and then look blank. Say, 'No — but I don't mind waiting, and I'd be pleased if you could ask for me.'

So, now's the big minute. Play it right, and — who knows? — you might even walk out of the door with a job. Fluff it up, and it's another chance down the drain. Imagine that you are the employer, busy at his day's work. As far as he is concerned, you are someone he has never met, who has come along unannounced and wants to take up his time. So if he begins to think that you are a nuisance, your chances will soon evaporate. On the other hand, if he *does* happen to have an unfilled vacancy, or if there is something about you that he likes, he will be willing to sit and talk to you because he will be wondering whether you would make a good employee — and taking on a new employee is a decision always worth a bit of effort.

When you get in to see him, shake his hand, and introduce yourself clearly. Explain that you don't want to take up his time unnecessarily, but that you are keen to find work with a good employer. Tell him a little about yourself. Hand him a copy of your PIC, saying, 'This will tell you something about me, if it's any help.' Once he has got that in his hand, most of the conversation should go to your advantage, since it will be about the things you chose to write on your PIC about yourself.

Soon after this stage, the conversation can go one of two ways — it can get more involved, which means that there may be some possibility of a job for you. Or he might say, 'Well, this is very interesting, but I'm afraid we haven't got any vacancies.' Don't just give up, and wilt out of the door — keep the initiative, and say 'Well, that's quite all right, I knew that I would be very lucky if you did have a vacancy just like that. But could I leave this copy of the PIC with you, and if something does come up in the future, could you consider me then?' This way you are leaving your 'visiting card' and your visit will not be a waste of time.

Leave with a smile and a handshake, and he might think, 'Well, she did seem to be a very pleasant and well-meaning person. I wonder if we *do* have anything coming up, where we could squeeze her in.' The impression you leave when you go will linger in the office for a while after you leave, and it is this that you will be remembered by.

There is a lot to be said for practice. Go over the things you might say with a friend (or with your support group, if you have one. See Chapter 5.). The shyest of people can learn to act as if they are confident, simply by practice. The more shy and unconfident you are, the more important it is to practise.

In conclusion: you are bound to meet some employers who are completely negative to you. Well, never mind. Just forget about them. The world is full of hard people, worried people and people without the time to talk to you. Don't get depressed. The world is also full of well-meaning, caring people who might be willing to go to some effort on your behalf if they take a liking to you.

7. The grapevine

Once upon a time, this was the *only* way that anyone ever heard about jobs. It is still a very important way.

- Do all your *friends, neighbours and relatives know* that you are down on your luck and looking for work? Do they know what *kind* of job you are looking for? Once they know, the word will be passed around.

  Make a list of all the people you know who might be able to help. Take a page in your Journal for this. Some will be more help than others, and have many more contacts. You can go back to them, every so often, to keep in close touch.

- *Contacts*. By getting involved in new activities in your neighbourhood, you will meet new people. You go to an evening class, and learn how to weld; you become friendly with a man in the class; he's got a friend who runs this garage; he mentions you to his friend; and you end up with a job in his garage. That is the way it works.

8. Local notice-boards

Local notice-boards such as you see outside newspaper shops and post offices often carry small advertisements for local jobs. For some kinds of work you can place a card in the window yourself. They only cost about 10p a week, and if you can make the card attractive in some way, it will stand out and be noticed. This is especially good for any kind of 'jack-of-all-trades' work; gardening, odd jobs, cleaning, etc. If you are not much good at design, ask a friend to help.

Some factories keep notice-boards outside with details of vacancies, but the habit is slowly dying out.

9. Local radio stations

Local radio stations often carry regular weekly spots covering news of jobs. Find out if your own radio station does so, and then tune in regularly. The BBC used to run an excellent weekly programme called 'World of Work', which has now stopped. Watch out for a new replacement programme.

10. Specialist Information Centres

Specialist *Information Centres* offer help and advice to people seeking work in their areas. See *Equal Opportunities: A Careers Guide for Women and Men*, by Ruth Miller (Penguin £1.95) for addresses.

11. Jobsharing

This is a method of working which is growing in popularity. Two people share one job. They decide between themselves and with the employer how they would like to divide the work (alternate weeks, 2½ days each a week, mornings and afternoons, etc.). Where employers have agreed to take on jobsharers, they generally like the arrangement because although they have a little extra administrative work to do, they gain two people's ideas and energy for the price of one.

The advantages for the people sharing the job are that they have that much more time to themselves — time to take another part-time job, time to work as self-employed or time just free. Jobsharing has been happening successfully for years in the USA and in Sweden, and it is becoming increasingly common in Britain.

In order to find a partner to share a job with, if you do not already know someone suitable, you could try advertising on local notice-boards or in the local papers. Perhaps in the future towns and cities may develop 'Jobsharing Banks' which could bring together other people wanting to share similar kinds of work so that they can make joint applications together. If the idea interests you, you can get a free leaflet called *Sharing a job?* if you send a stamped addressed envelope to Adrienne Boyle, The Job-Sharing Project, 75 Balfour Street, London SE17.

A jobhunting checklist

To help you tackle your jobhunting efficiently, Figure 8 is a chart on which you can keep a weekly check on your activities.

Each week, tick off the things that you have done, and tot up your 'score'. It is a very primitive method of scoring, so do feel free to change it. All that really matters is that you use the same system of scoring each week, as this will then give you some idea how you are doing.

What if you find that you are scoring less, week by week?

- You can tear the page out, so that you no longer have any evidence that your efforts are slipping.
- You can ask yourself why you are slipping.
- You can take a full week or two's holiday from jobhunting, to give yourself a break, and then return to it with a fresh spirit.
- You might be chasing the wrong kind of work? Maybe Chapter 3 may help.
- You might simply have set a very high standard at the beginning, and it doesn't matter if you fall off a bit, as long as you don't give up altogether. Chapter 2 has ideas about survival when the going gets tough.

Week beginning	Date	Date	Date	Date	Date
Visited/phoned the Careers Office					
Visited/phoned the Jobcentre					
Visited/phoned the employment agencies How many?					
Checked the papers/journals? How many?					
Written letters 'on spec'? How many?					
Went out visiting. How many employers?					
Talked to friends or neighbours about your jobhunting? How many?					
Met any new people? How many?					
Checked local notice-boards?					
Listened to local radio jobhunting slot?					
Sent off application letters? How many?					
Had any replies? How many?					
TOTAL SCORE:					

Fig. 8. Jobhunting checklist.

STAGE 3: APPLICATION LETTERS AND FORMS

Letters

Put yourself in the position of the employer. She (or he) has just advertised a vacancy, and the letters start pouring in. There may be 20 or 200 letters of application. Out of all of these, she has to pick just 5 or 10 to ask to an interview. What would *you* do? How would *you* pick out the small handful which 'win' this stage of the competition, and get invited to the grand finale at Company House? We will look at ways in which you can try to get your own letter among the winners.

1. *The first throw-out:* to begin with, she does not think about the good letters. Her first interest is in *the bad ones* — the ones she can throw straight into the bin. She will probably throw the letters out in two stages. Firstly, she will throw out the letters that are dirty and badly written. Remember, she *wants* to find an excuse for throwing all but a handful away, so any excuse will be pounced on, and scrunch, toss, all your hopes lie crumpled in that sad place, the rubbish bin.

2. *The second throw-out:* she will then go through the letters again, and toss out a further batch, until she has about twice the number that she wants to interview left. At this stage, she will only keep the letters which:

- Are not too long.
- Give relevant information about the applicant.
- Avoid any attempt to show off.
- Don't waffle, or give unnecessary information.

3. *The final choice:* finally, she will go through the pile a third time, and this time she is looking for something positive, that certain 'something' that says, 'This person seems to be good. He (or she) seems to be the kind of person we ought to ask to an interview.'

Preparations
Check the advertisement carefully. Phone up for more information if you need to. (See above.) As well as the stated requirements there may also be 'hidden' requirements. Make two lists, one of the stated, and one of the hidden requirements, as in Figure 9.

MOTORCYCLE MESSENGER required

with knowledge of city

and clean licence

Stated requirements

Ability to ride motorcycle
knowledge of the city
clean licence

Hidden requirements

reliability
responsibility

Fig. 9. Sample ad.

'The Old Scrapyard',
Dungheap Lane,
Dirty Old Town, (1)
Mouldshire.

Tel: D.O.T. 3898

June 15th 1982 (2)

(3) The Personnel Manager,
Super Smart Ltd.,
Posh Street,
Snootyville.

(4) Dear Madam or Sir,

re: Position of General Cleaner (5)

(6) I would like to apply for the above post, as advertised in the Snootyville News on June 13th.

I have had several years' experience of general cleaning and maintenance work, which has most recently included a year with Everlasting Filth Ltd., where I held the position of Supervisor, responsible for 20 part-time sweepers and one mechanical sludge-eater. I enjoyed the ordinary work, as well as the tasks involved in supervision. I would be perfectly happy to accept an ordinary position as General Cleaner, as I find the work pleasurable, and enjoy doing a good job. (7) (8)

In a previous job with Awful Boilers Ltd., I trained as a specialist in cleaning uncleanable boilers, and also acquired a general skill at mending faulty taps, valves and pipes. (7) (8)

I hope to hear from you soon.

Yours faithfully,
Adam Edensweeper
(ADAM EDENSWEEPER) (10)

Fig. 10. Sample letter

1. Follow this general layout. Your own address at the top, with a phone number. (If you are not on the phone, see Figure 5.)
2. The date.
3. The address of the firm.
4. If you know the name of the person you are writing to, this is best. You might phone the receptionist, to ask?

5. Clear indication of the post you are applying for.

6. Clear opening sentence.

7. Details of previous experience and employment.

8. Show that you can meet the stated and hidden requirements of the job. (Don't apologise for experience or qualifications you *haven't* got.)

9. Keep it short. If it's too long, employers might not read it at all. Avoid unnecessary details — hobbies, religion, previous salary, reasons for leaving the last job, etc. Only put in things that would interest *you*, if you were the employer. One side of paper is enough (if you are using large A4 paper). You can always include your PIC or c.v. to give extra information.

10. Print your name underneath your signature.

Make sure the whole letter is clean, clear and tidy, without any crossings out or mistakes. This is why it is always best to do a rough version first. Try to avoid any stale or over-used phrases.

Trying to write a good application letter

After you have written it.
  ● Now, the surprise — don't post it! Put it away for a day or two. When you come back to it with a fresh mind, you may be able to improve it. The employer won't fill the post all at once, so you'll lose nothing.
  ● Post it first class in a clean white envelope, clearly addressed. Keep a copy for yourself, so that you know what you said if you are asked to an interview.
  ● Then, wait. If you don't hear after ten days, telephone them and ask, 'I just wanted to check if my application letter for such-and-such a post had arrived.' This will put your mind at rest, and it won't reduce your chance

(unless you say, 'Have I got the job?' or something equally pushy). Above all, it will stop you getting very angry with the firm, and — who knows? — you might well make a good impression!

Don't be thrown in the bin!!
Use this checklist each time you write a letter of application. Do not post your letter unless it has a tick by *every* point.

First hurdle	Clean sheet of paper, not lined.	☐
	Straight margin, straight lines.	☐
	Clean and tidy appearance, no crossings out.	☐
	Correct spelling and grammar.	☐
	Clean envelope, 1st class.	☐
Second hurdle	Short and to the point. No waffle.	☐
	Orderly presentation.	☐
	Relevant details about your experience and qualifications.	☐
	No additional personal details.	☐
	No mention of previous salary.	☐
	No mention of why you left last job.	☐
	Something to suggest you have noticed and can meet the 'hidden requirements'.	☐
Third hurdle	A calm and confident impression.	☐
	Something to imply that you are hard-working, responsible, etc.	☐
	Something to make the employer think, 'I must meet this person, to find out more.'	☐
	Two days delay before posting.	☐

Spelling hints:

Personnel	Advertisement	Assistant
Experience	Responsible	Opportunity
Sincerely	Appropriate	Requirements
Faithfully	Relevant	Dictionary (for the rest)

For further advice: Howard Dowding and Sheila Boyce's book *Getting the Job You Want* (Ward Lock £1.50) has more detailed advice on letter-writing, interviews and other aspects of jobhunting. (Library, if you can't afford to buy it.)

The application form

Nearly every person who is looking for a job has to fill in an application form sooner or later. They are fairly straightforward things, but it is easy to make mistakes. Care and practice can help to make sure that yours does not bite the dust unnecessarily.

How can you practise?

The best way to practise is to get hold of a form. Simply phone up any firm which is advertising vacancies in a type of work which might interest you, and ask if they would send you an application form. If you don't strike lucky first time, try again.

Here are a few guidelines, to help you fill it in:

- Fill it in in pencil first. Ask a friend to check it over, before you finally fill it in in ink. It *always* helps to have a second person look at your form, letter, PIC, c.v., etc.
- Follow the instructions very carefully. Keep your form neat and tidy. Don't overflow the spaces. If you *do* need to overflow, use any blank space that is provided, or the back, or an extra piece of blank paper.
- Under 'Previous employment', you might be asked to say why you left your last job. Be careful to write something that sounds good, such as, 'I wasn't satisfied with the work' or, 'There was no opportunity for training' or, if you were sacked, 'Unsatisfactory relationship with the employer'.
- Under 'Position and duties' describe what kind of work you did, and responsibilities that you held.
- Under 'Hobbies and interests', concentrate on the ones that seem relevant to the job, which might help you to get it. If you have no interests that you think worth writing down, it might be an idea to read through Chapter 8 and to take up some new interests, since they can help you in unexpected ways — e.g. getting experience, making new contacts, getting references, learning new skills.
- Under 'Any other information' you have a chance to mention anything else which you think might help you to get the job. You might be able to bring in some of the 'achievements' from Chapter 3.
- Under 'Referees', give the names and addresses of two people who will speak well of you, and whom an employer would respect.

When you have completed your practice form, keep it somewhere safe. The next time you have to fill one in, you will find it much easier. Alternatively, you might decide to post it off to the employer you got it from, in answer to the original advertisement you saw! If you do this, make a copy of what you have written, for future use.

- When posting an application form, send a short covering letter with it. This makes it personal, and leaves a very good impression. Check that your letter is neat and tidy.
- If you haven't heard back within 10 days, telephone to see what is happening.

STAGE 4: THE INTERVIEW

If you have got as far as having an interview, you are almost home, so it is worth putting a lot of care into preparations. Put yourself in the position of the interviewer. She has to choose from five or ten people. *What is she looking for?* She will have three kinds of question in her mind when she meets you, and by thinking carefully about each one, you can prepare yourself.

1. She wants to satisfy herself that you are *basically polite, tidy and pleasant*. She wants to be sure that the person she chooses will fit into the firm, office, team, or whatever, and not be a constant source of difficulty.

2. She wants to find out either if you have got the *skills and experience* for the job, or if you seem likely to learn. It is not just qualifications that the interviewer is looking for; there are a lot of other, more ordinary, human qualities that she will be pleased to see. If she is impressed by a person, and starts thinking, 'He would be a good person to have around', then she might not bother so much about the lack of the right paper qualifications.

   You should remember that there are many ways of getting 'experience', apart from actually *doing* the work in question. Say you are applying for a job in an accounts office, and have never done accounts work before. This need not matter — what about that job you held as an assistant in a library, when you had to keep the books in exact order? Isn't that the same kind of skill? You had to be tidy and accurate. You must think *around* the job. Try to work out what *general* kinds of ability are required, and then think back (Chapter 3 will help you there) to see whether anything you have done has shown this general kind of ability.

   Then, when the discussion at the interview comes around to the matter of your experience, you can say, 'Well, I have never worked in an accounts office before, but I have thought about it, and it seems to me that a person working in an accounts office needs to be tidy and accurate. This is something I enjoy doing. I think that my experience in the library, where I had to check the cataloguing and keep the books in order on the shelves, shows that I have this kind of ability. I hope I can pick up the details of accounts work quite quickly.'

   What if you haven't got the right exams or qualifications? Assuming that this doesn't rule you out automatically, you could say that you would like to work for the necessary qualifications during the year, at evening class, if that would be all right. In other words — *don't assume defeat until you are actually defeated*. There is always more than just one way of doing something.

3. She will be looking for *sincerity*. She wants someone who is going to be serious about the job, someone who will be honest, and reliable. Then as well as these qualities, she will be *hoping* to find someone who shows this sincerity in a *positive spirit* towards the job, someone who has shown sufficient interest in the job to have done some homework on it before the interview.

Preparations for the interview

1. *Finding out information about the firm, or job*
You could ask for copies of the firm's literature — leaflets, annual report, etc. If

you phone the firm in advance, they might send you some stuff in the post. You could also ask around among your friends, and you could ask for information at the Careers Office and the Jobcentre.

2. *Thinking ahead about the job*
What does the job need? What are its stated and its hidden requirements? What kind of experience have you had, which you could talk about? Why do *you* think you could do the job well?

3. *Thinking ahead about questions and answers*
- 'Why do you want this job?'
- 'How much do you already know about the job?'
- 'Do you think you would be able to manage the heavy demands of the job?'

Each question gives you an opportunity to talk. Don't just answer 'Yes' or 'No' to questions. Have courage, even if you haven't got much confidence!
- Finally, 'Do you have any questions you want to ask?'

You should prepare for this last question. You might want to ask about training, or details of the job. Don't start asking about pay until you have asked the questions which show that you are interested in the job. It is quite OK for you to take a notebook in with you, with a short list of questions you want to ask. It will help, because nerves very often make us forget everything we had planned to say. If you can't think of anything to ask, just pause, and then say, 'No, I don't think so, thank you.'

4. *Practice in interview technique*
- If you have either joined or formed a support group, this is the kind of thing that is good to do together. You can do 'dummy' interviews, and practise what you might say with your friends. A local employer might come in to give you a hand, and add a touch of reality. (See Chapter 5.)
- If you are on your own, you can ask a friend to come and help, by giving you a 'dummy' interview, perhaps with a third person present, to make it seem serious and real.
- If you don't know anyone who you think could help, you might ask at the Careers Office if they would help you, or if they have any suggestions. They might be willing to get a small group together for a practice session. Another thing that helps is practice in 'public speaking'. Ask a friend if he or she would sit, listen and then comment, while you give a short one minute (or three minute) talk, on any subject you like. After you have done this four or five times you should find your confidence and your manner improving a lot.

5. *Practical preparations*
- *Appearance:* It is worth putting an effort into your appearance, for this

one day, so that you are well-dressed and tidy. Prepare the night before.

- *Getting there:* plan your journey, and work out how long it will take you to get there. Plan to get there about 15 minutes early. If you are travelling by bus, work out the bus times the day before — don't leave it until the morning to find out that they cancelled the 63 bus months ago.

  If you *are* going to be late, phone up to say so, and then it will probably be all right.

- *What should you take with you?*   *Tick*

  The details of the interview: time, place, name of the person seeing
  you, phone number.   ☐☐☐☐

  5p pieces for the phone, in case you are late.   ☐

  Your PIC or c.v.   ☐

  Details of your previous employment and NI number, in case you have
  to fill in an application form.   ☐☐

  Pencil and rubber, for the application form (and a sheet of spare
  paper).   ☐☐☐

  Any notes you have made for the interview, questions you want to
  ask, etc.   ☐

  A 'memory card' (see end of this chapter).   ☐

  A watch, a clean hanky, a smile.   ☐☐☐

This last thing is important. Take a positive attitude with you, because it will show, and the interviewer will like it.

The big day

Feeling nervous? Don't worry — it happens to everyone. Your body is creating adrenalin. Think of it as extra energy that you have got, which you can actually use in the interview.

The first few minutes: knock confidently, walk in confidently. If you've got a handshake like a sloppy wet fish, then for your own sake practise a good handshake beforehand. Don't smoke, unless invited to. Sit back comfortably in your chair; don't perch nervously on the edge.

The interview will probably (but not certainly) be run along these lines:

1. Telling you something about the job and the firm (or organisation).
2. Asking you questions designed to find out how well you would fit the job. (This is your chance to talk about experiences you have had which you think are relevant to the job).

Feeling nervous

3. Inviting you to ask any questions.

Should you use your PIC? If you think that it says useful things about you which you might not get a chance to talk about, then yes, show it to the interviewer. On the whole it is better if you can talk about yourself directly, and say why you think you can do the job. If you are less confident, your PIC might be a great help.

When you are talking, *stick to the point*. Don't gabble off on all sorts of things — that is another common mistake people make when they are nervous. When you are asked a question, *pause and think* before answering. There is no hurry.

If you don't succeed . . . just keep on trying. You can learn from your failures if you sit down and think about each interview after you have done it. Make some notes about the interview in your Journal as soon as you get home (before you even know the results), and note down any reminders for next time, e.g. 'I *must* think before answering questions.'

Unless you are very lucky, it *will* take you several interviews before you land a job. Treat the interviews as opportunities for practice — then you won't be so disappointed, or think that they were just wasted time.

Professional and executive interviews
You may sometimes have to attend 2 or 3 interviews, perhaps a panel interview, and perhaps undergo a group exercise or test. The book *Getting the Job You Want* by Howard Dowding and Sheila Boyce (Ward Lock £1.50) has advice on this kind of interview.

Interview memory card

Copy these 9 points on to a postcard, and take it with you to interviews as a reminder:

	Check before	Check after — did I ?
1. Have I got everything I need to take with me?	☐	☐
2. Have I checked the travel times? (15 minutes early)	☐	☐
3. Clean and tidy?	☐	☐
4. Knock, enter, smile, shake hands confidently.	☐	☐
5. Pause, before answering.	☐	☐
6. Don't just answer 'Yes' or 'No'.	☐	☐
7. Stick to the point.	☐	☐
8. Have courage.	☐	☐
9. Make sure I get a chance to say the things I want to say.	☐	☐

TELEPHONE TECHNIQUE

If you are not very confident on the telephone, you ought to think about getting some practice. Ask someone you know if she would agree to pretend to be an employer who is advertising a job, and then phone up to ask for information about it. Practise both from an ordinary phone, and from a call-box. When you are more confident, start getting 'real' practice by phoning up for real jobs advertised in the paper. Practice is the only way to build up confidence.

Here are some tips:

- Always start with 'Hello'. Then give your name.
- Explain clearly why you are telephoning, e.g. 'I am phoning for more information about the assistant's job you are advertising in the News' or, 'I am due in for an interview at 10.00 with Mr Miller, and I am phoning to say that I will be ten minutes late.'
- If someone says, 'Hold the line, I'll put you through' always start again from the beginning with the new person. It can sometimes take a while to get 'put through', so be patient.
- If you can find out the name of the manager or personnel manager from the receptionist before you can speak to him, this will help you, since you can then say, 'Mr Edwards?' or, 'Can I speak to Mr Edwards?'
- Always have a sheet of paper and a pen handy when you telephone, ready to take notes. A list to remind you what you want is also helpful.

Finally: if possible, arrange to make your call from an ordinary phone. If you do have to use a coin-box, be sure to have plenty of 5p pieces with you, and ask to be called back if your money runs out.

OBSTACLES TO JOBHUNTING

Racial discrimination

Do you apply for jobs, and find that you are always being told, 'Oh, we're so sorry, the job's been filled'? If you suspect it's because you don't possess the regulation white skin, what can you do?

If you suspect discrimination, arrange with a white friend that you and he or she both go along for the same job, one after the other. If your friend is offered the job just after you have been told it has been taken, then you've got good evidence. But be prepared — the employer will fight!

You can take your case to an industrial tribunal if you can get good evidence that an employer is discriminating against you because of your race, colour or nationality. This also applies if you have evidence that you have been offered a job on less favourable terms than others, because of the colour of your skin.

Before taking any definite action against the employer, you might do well to consult with others who have experience of this kind of thing — because it isn't easy to make the evidence stick, however blatant it seems to you. A youth or community worker, a Neighbourhood Law Centre or the Citizens Advice Bureau will help you. You can get free legal advice under the 'Green Form' scheme — but your chance of success might depend on finding a really sympathetic solicitor who is willing to take an interest in your case.

Citizens Advice Bureaux and Jobcentres both have copies of a pamphlet called *Racial Discrimination*, which explains how to bring a case before a

Tribunal.

For advice and support write to the Commission for Racial Equality, Elliot House, 10/12 Allington Street, London SW1E 5EH (01-828 7022).

Sex discrimination

The same advice applies as above. An employer offering a job must not discriminate against you on the grounds of your sex, and nor must he offer you a job on less favourable terms. (There are a few exceptions, such as work in single-sex establishments, or in private households.)

If you think you have been discriminated against, *do* have the confidence to do something about it. For advice and support, you should write to the Equal Opportunities Commission, Overseas House, Quay Street, Manchester M3 3HN (061-833 9244).

You are having special problems in getting a job because you have a criminal record

The 1975 'Rehabilitation of Offenders' Act sets out the periods of time after which certain types of conviction become 'spent'. This means that you do not have to reveal the details of any such 'spent' convictions when you are applying for a job, or filling in a form. You can get a leaflet from the Jobcentre/Employment Office called *Wiping the Slate Clean*, which gives all the details.

In general, you can go a long way towards wiping your own slate clean by the kinds of things you decide to do while you are unemployed. In an employer's eye, if you have a prison record but have been actively helping out in the neighbourhood with any kind of project or scheme, this will count in your favour whatever your past may be, and he may decide that your present activities are more important than your past ones when considering you for the job.

Young people with a record should ask about the Youth Opportunities Programme, especially the training workshop and community service schemes, which usually include special help for young people with special problems, and which often have helpful sympathetic people working for them.

Transport problems

Getting around, to apply for jobs, etc: don't forget the good old bicycle. It's often quicker than the bus, often quicker than the car (you can sail past traffic jams), it's free and it keeps you fit.

When it comes to getting to a job, the bicycle may again be ideal. You might also be able to share a lift with other people going that way — you could advertise locally, if there's no one in the job who lives nearby. You can buy a secondhand moped for around £150, and they are very cheap to run and insure.

Chapter 5: Support groups

WHAT IS THE VALUE OF A SUPPORT GROUP? — HOW CAN YOU SET UP A GROUP? — THE FIRST SESSION — MEETING NEEDS: WORK THAT THE GROUP CAN DO — FURTHER ACTIVITIES — GENERAL SUGGESTIONS — GETTING HELP

A support group

WHAT IS THE VALUE OF A SUPPORT GROUP?

A problem shared is a problem lighter. If people who are unemployed can find a way to meet together regularly to help each other, a new source of strength and encouragement can be discovered.

This chapter is intended to offer some practical ideas and to suggest some structures and guidelines for a self-help support group. There are various kinds of self-help groups meeting across the country, but very few unemployment self-help groups yet, so there is little accumulation of experience to go by. If you set up a group, you may find these notes helpful — but bear in mind, also, that you are exploring new 'territory', and that you may want to develop new approaches for your support group as you go along.

HOW CAN YOU SET UP A GROUP?

1. Talk among your friends and colleagues; show them this Handbook, and plan a first meeting to discuss the idea. If two or three people jointly decide to

launch a support group and attract other people to join it, this provides a better basis than if just one person starts the initiative. There is often a sense of dependence on the person or people who start an initiative, and people who join later tend to look to them for leadership.

2. Advertise locally. Use cards on newsagents' notice-boards, an article in the press, leaflets given out in the queues on signing-on day, posters around the place.

   Seven is a good number for a group — but since there are always one or two who cannot turn up, plan on having 12 members. If it gets above 12 at any meeting, some people may not get a chance to join in. The best advice, if a group grows above 12, is to separate into two groups.

THE FIRST SESSION

1. Introduce yourselves around the group. Tell each other about your background, your previous job(s), how you come to be out of work, and anything else that might be interesting.

   One of the important ingredients of a group's success is sharing and listening. Each person must feel that he or she has a chance to talk. The rest of the group must listen without interrupting if everyone is to feel an equal member. This also helps people to talk more clearly.

   If this guideline is not followed, some people will talk a lot, some may not talk at all, and dissatisfaction may develop.

2. Share your *feelings* about being unemployed. Go round the group so that everyone has a turn. If some people talk and talk and talk, you should agree to have a two or three minute time rule, so that everyone gets a turn. Some people may have feelings which trouble them; if they can talk about them, they can discover that they are not alone and that they have support and understanding for their feelings. 'Sharing' like this is valuable, and tends to create a sensitive, caring mood in the group.

3. Discuss what your needs are. Each person may have different needs, and it may not be easy to plan group sessions so that everyone's needs are met.

   Using the list of 'Meeting Needs' (Figure 11), take five minutes which each person goes through the list and ticks off the points which matter to him or her (using Column A). Then take it in turn to read out your lists, while the rest of the group keeps a collective score for the group as a whole, in Column B. This will help you all to see which needs are shared by all, which are particular to just one or two people, and will help you to plan how you spend your time at meetings.

		Yourself Col. A	The Group Col. B	Total
1.	Would like help with jobhunting.	☐	☐	....
2.	Would like help with application letters.	☐	☐	....
3.	Would like help with interview technique.	☐	☐	....
4.	Would like help with telephone technique.	☐	☐	....
5.	Would like help with writing a PIC or c.v.	☐	☐	....
6.	Uncertain what to do, and what kind of job, training, etc. to seek.	☐	☐	....
7.	Would like people to talk to, about some of the problems of being unemployed, and about things in general.	☐	☐	....
8.	Have difficulty in keeping up a regular effort each week.	☐	☐	....
9.	Boredom — don't know how to fill up the time.	☐	☐	....
10.	Need to have some fun.	☐	☐	....
11.	Problems over benefits, or other money problems.	☐	☐	....
12.	Would like to explore the possibilities of self-employment.	☐	☐	....

Fig. 11. Meeting needs checklist.

MEETING NEEDS: WORK THAT THE GROUP CAN DO

1. *Jobhunting:* exchange experiences. Go through Chapter 4 discussing each method mentioned, swapping experiences and suggestions. What works? What doesn't work? Write down any decisions that you make and make sure that you will try to do them during the week. Each person could make a point of deciding to do something they haven't done before. It is a good idea to have a definite time in the following meeting to ask everyone about how they got on. Discuss how you can help each other to find work.

2. *Application letters:* go through the section in Chapter 4 together. Write sample letters in the group and discuss each other's letters.

3. *Interview technique:* you could practise together by setting up a 'role-play'. Two people act the part of interviewers, and the others take it in turn to be interviewed. Prepare for this in advance by deciding what the job will be, what the terms are, what questions the interviewers will ask, etc. Afterwards, discuss how you did, and how you felt, what your fears were, etc. Give

support to people who are obviously nervous.

A local employer might agree to come in and help, and run a dummy interview session with you. The local college might agree to let you use their video equipment so that you can film yourselves and then play it back afterwards to watch yourselves. Get in touch with someone at the college who is responsible for 'Life and Social Skills' courses for young people on the 'Youth Opportunities Programme' who will understand what your needs are. If you cannot locate anyone at the college, try asking at the Careers Office. A drama teacher either from a school or the college might be able to help by setting up an evening devoted to role-play interviews. (Role-plays are both very instructive, and fun.) An ordinary tape recorder is helpful too. Just tape the interview and then listen to it afterwards.

4.   *Help with telephone technique:* arrange to practise together. (See Chapter 4.)

5.   *PIC or c.v.:* use the notes in Chapter 4, and do this together one session. Make an arrangement for photocopying.

6.   *Uncertain what to do:* work through the exercises in Chapter 3 together. (This could take up to two hours, with discussion.)

7.   *Need people to talk to about things:* arrange to spend some time in your group each week in pairs, so that people can talk to each other on a one-to-one basis, taking it in turns to talk and then to listen. If some people are not comfortable working in twos, the group could break into threes.

8.   *Keeping up a regular effort:* make weekly commitments to the group of what you plan to do each week. *Write them down.* Then report back each week on how you got on. It would be good to put time aside each week so that each person could have three minutes to say what he or she has done during the week. Give each other help and encouragement.

9.   *Boredom:* work through Chapter 8 together. Discuss the kind of things you like doing in leisure time. Are you still able to do them? What would you like to do now, if you had the energy or enthusiasm? Let each person in the group have a chance to have a say. Think of something you have never done before which you would like to try, and then find out if it is possible using advice in Chapter 8 and the telephone. (Gliding? Parachuting? Rock-climbing? Roller-skating? Hiking for a weekend? Sauna? Photography? Stripping down a motor-bike? Working with a housing group? Organising a march, or meeting, about local unemployment?) Make a commitment to the group about what you plan to do during the coming week, and then report back to the group on how you did at the next meeting. Two or three people might agree to do something together.

10.   *Need to have some fun:* take a day out or an evening out together. Go to a film/dance/pub/beach together. Throw a party — each person could cook and bring something. Take the children all out together.

11.   *Problems over benefits:* use Chapter 10 to go through each person's prob-

lems. Put aside some time every time you meet in case new problems have arisen since last week. You could accompany each other to tribunals, too. Obtain a copy of the *National Welfare Benefits Handbook* (see Chapter 10.).
Other money problems: discuss them, giving each person a turn, and making suggestions. Ideas in Chapters 9 and 10 might help. Some problems could be solved within the group, such as lending things to each other (e.g. tools, books, children's clothes, equipment, etc). You might be able to help paint each other's houses, dig each other's gardens, etc. Chapter 10 has advice about how much money you are allowed to earn without breaking the benefit earnings rules.

12. *Self-employment.* Work through Chapter 9 together, and discuss your ideas. If one or two people want to pursue this, they could try to find a few other people and then set up a self-employment self-help group as suggested in Chapter 9.

FURTHER ACTIVITIES

A group could organise a joint visit to the local college, or to the 'TOPS' Skillcentre, to see what it is like there and to find out more about courses. You could have sessions when someone came to speak to you, perhaps from the DHSS or the Unemployment Benefit Office, to discuss each other's problems.

You might feel that you want to look at the wider problem of unemployment in your area, and start asking some relevant questions. There is a lot that a group of people working together can do, once they have agreed that they feel strongly about something and want to start doing something. Ideas and suggestions in Chapter 12. You could study one or two of the books mentioned in that chapter, and discuss them, as you worked out how your own neighbourhood might best respond to its own local problems. One good answer to a problem is often to get together with other people sharing the same problem, and to begin to do something about it together.

GENERAL SUGGESTIONS

- Make sure that you share responsibility for running and organising the groups — don't leave it all to one person.
- Plan next week's session each week — don't leave things too loose, or the group may begin to disintegrate. Regular attendance is quite important, so that you can begin to build up trust with the others in the group.
- At the outset it might be a good idea to set a time limit— that you all agree to try to come to a series of (e.g.) 8 meetings, after which you will decide whether to continue. People should agree to keep on coming, even if they get a job in the meantime (if they are still able to). For this reason, it would be good to meet in the evening.
- If you have practical arrangements to make for the next session (e.g. for interview practice), make sure that you know who has taken on the re-

sponsibility to do what. Don't leave it vague.

- Talking — be sure to leave space for the quieter members who may not be so confident about speaking in front of the whole group. Try to be aware who the busy talkers are, who has a habit of interrupting and why they do it.

- Express your feelings, as well as your ideas and your actions. Bringing in feelings helps to bind a group together, because you are offering more of yourself to the others, which has an 'opening up' effect on the whole group. Be willing to bring your personal life into the group.

- *Every week* allocate some time so that each person present can have 2 or 3 minutes to talk about his or her week. Ask each person to tell the group about something gloomy and then something positive and good that happened. Just listen, without interruption.

  Allocate a definite time for each person to talk about jobhunting progress, or lack of progress. The more you share of each other's problems, the more you will be able to help each other. Some groups may want to share a lot, whereas others may not — this is fine. Don't feel pressured into sharing things, if you feel that you don't know or trust the group enough. (Contradictory advice — that's life!)

- *When and where to meet* — a regular time is a good idea, so that group members can plan their other activities around it. You can either meet in the same place each week, or take it in turns to visit each other's homes, which is also pleasant.

GETTING HELP
In most areas, there ought to be *someone* who has experience in this kind of group work, who might be willing to come and give you some help or advice. How can you find whoever it might be?

- Find out if there is a 'YOP' scheme in your area. YOP schemes involve team-work, and the staff often have group work skills. Ask through the Careers Office, and then get in touch with the project direct.

- The local college. If there is a YOP scheme, young people might be going to college for one day a week, and the staff taking their sessions at college may be helpful. Enquire at the college, or via the Careers Office.

- The local youth service might be able to help (for groups of all ages).

- The Samaritans have trained counsellors among their volunteers. Contact their local office, to see if they can help.

- Write to *New Society*, 30 Southampton Street, London WC2, telling the editor about your group's work, and asking if they could print a little note about your needs.

- Do the same for the magazine *Youth in Society*, 17 Albion Street, Leicester, which is read by many people who have skills in group work.

- The BBC community self-help programme *Grapevine* might be interested to hear about your group, and its problems: *Grapevine*, BBC TV, London W12 8QT.

● There are a few more notes and ideas about the value of support groups in Chapter 12.

---

1. Listen without interruption when someone is talking.

2. Time people's contributions — e.g. 'Let's give everyone three minutes to say what happened to them last week.'

3. Be aware of the other people in the group.

4. Be honest about your own feelings.

5. Trust the other people. Ask for confidentiality about what you say, if you want it.

6. If you are the leader and a conflict breaks out, in the first place let the other members of the group deal with it. This is the best way of leading (from behind).

7. If you are leader don't try to be perfect. Remember that you are also just part of a group of people with the same needs and aims.

8. Trust to a group atmosphere which can often work in a 'magic' way, if you are honest and care for each other.

9. Enjoy being in the group. If you make some contribution to it, you will get a lot out of it.

10. Don't be judgemental. You may often find that if you seem to be judging others it is yourself that you are talking about, not the other person.

Fig. 12. Some guidelines for groups.

Chapter 6: Special notes

6A: NOTES FOR YOUNG PEOPLE – 6B: NOTES FOR WOMEN – 6C: NOTES FOR DISABLED PEOPLE – 6D: EARLY RETIREMENT: NOTES FOR OLDER PEOPLE – 6E: NOTES FOR REDUNDANT WORKERS – 6F: NOTES FOR PROFESSIONAL AND EXECUTIVE WORKERS

This chapter gives extra information of particular value to the people listed above.

Chapter 6A: Notes for young people

MAKE AN ORGANISED EFFORT TO GET A JOB – GET A PLACE ON THE YOUTH OPPORTUNITIES PROGRAMME – GET A PLACE AT COLLEGE – IF YOU ARE COMING UP TO 19, GET A PLACE TO TRAIN ON A TOPS COURSE – ACTIVE UNEMPLOYMENT: DO IT YOURSELF – FORM A SELF-HELP GROUP WITH OTHER YOUNG UNEMPLOYED PEOPLE.

Join the game. And what a game!

The key to your problems is *persistence*. Don't give up. The system has let you down, and there aren't enough jobs for everyone, but don't let this get you down. Don't stop believing in yourself. Hell, no! If you feel angry – then shout. If you want to do something – then do something. If you can look at things positively, believe in yourself and stand up for yourself, you can still get what you want.

There are probably more things that you can do than you have thought of. This chapter takes you through these different possibilities. Read about them, and then decide what you are going to put your efforts into. Make a decision, make a plan, and then *persist*.

Before starting

There are some basic things that you may find helpful.

1. The Basic Skills Unit at the National Extension College has produced a wide range of materials for young people without jobs, including:
   - Help with reading, learning, maths, etc.
   - Booklets written by other young people.
   - Guide to working in shops, garages, etc.

   Write to BSU (c/o NEC, 18 Brooklands Avenue, Cambridge CB2 2HN) for information.

2. *School-leaver — this is your life*. Basic information on work, sex, housing, money, police. It has been produced by people in Tottenham, and all the addresses are for Tottenham, but it's still very helpful. If you set up a self-help group you could reprint it with your own local addresses. Send 50p to the Tottenham Neighbourhood Law Centre (15 West Green Road, London N15). (£1 for adults, please.)

3. The *Action Special* series (run recently on BBC Radio 1) produced, with MSC help, a number of fact sheets on particular jobs. If you write to *Action Special* (PO Box 101, London E1 9NE) stating the job(s) that interest you, they may be able to help.

OK. Now for the six possibilities.

1. MAKE AN ORGANISED EFFORT TO GET A JOB

Most of the stuff you need is in Chapter 4 — or in Chapter 3, if you don't know what kind of job to look for — so we will not reprint it here. All the key points about places where you can look for work, writing a Personal Information Chart, writing a good application letter, going for an interview, etc. are there. The only thing that is missing are any proper notes about the Careers Service, so we'll give those here.

The Careers Service

If you are under 19 you can choose to register for work with the Careers Office or the Jobcentre (see Chapter 4) and you can use both to find jobs. But you still have to sign on fortnightly at the Unemployment Benefit Office. The Careers Office is there to help you and has trained staff. Their address is on page vii.

To get the most out of the Careers Office, there are some basic rules that you should follow:
- Make an appointment by phone if you want to see a Careers Officer, so that you are sure of getting proper attention.
- When you call in for your appointment, take some ideas in with you about the kind of work you'd like to do. It makes life much easier for the Careers Officers. Use Chapter 3, if you don't know what kind of work you want to do. This way, too, you can avoid being pushed into a job you don't like.
- Keep in touch regularly, to let them know that you are still interested.

When there are not many jobs to offer, it is only natural that the staff tend to offer interviews to the young people who show the most interest.

- If they send you for an interview for a job you find you don't want, don't give up on them. Go back, and tell them why you didn't want it.
- The Careers Office staff may also be able to give you some helpful comments about your application letters, if you show them a copy of what you write. They *may* be able to arrange an interview practice session too, if you ask.
- Finally, remember they are there to help you, but this doesn't mean that you needn't do anything to help youself. Make your own efforts too.

2. GET A PLACE ON THE YOUTH OPPORTUNITIES PROGRAMME (YOP)

The Youth Opportunities Programme (YOP) has been set up by the government to give young people a chance to get the experience that all the employers say they want. If you are unemployed and under 19 you can join one of the YOP schemes and get paid a weekly allowance by the government. It's not much, we know, but look at it this way: it's more than the dole is, and the months that you spend with YOP might help you to get a 'proper' job and wage when you finish. And if you don't do it? Well, if you remain unemployed, not only will you get less cash, but you also won't get any experience, which will set you right back when you come to apply for jobs (unless you choose to follow one of the other possibilities 3—5).

Some young people say, 'It's not a proper job' or, 'It doesn't pay enough'; but it might set you up for something good later. And what's more, with 'Work Experience' you can get a chance to work at the kind of thing you *want* to work at. Garage mechanics? You can get work experience at a garage. Hairdressing? You can try out a hairdresser's. Working in a theatre? Get work experience with a theatre company. How? The Careers Office can try to arrange it for you. Anyway, here are the details about YOP:

There are several different YOP schemes, as follows:

- *Work Experience* (*on employers' premises*). This is the biggest scheme within YOP. It gives you a chance to work with an employer for 6 months, so you can try your hand at various different kinds of work while you are there.
- *Short Training Courses* covering construction, catering, engineering, vehicle servicing, retail distribution, office skills, etc. The courses last about 12 weeks, and give you just a basic introduction to the skill (not a thorough training, which is impossible in 12 weeks).
- *Project-based work experience, community service and training workshops*. The first two of these involve particular projects and schemes that are of value to the community — they might involve work with children, or they might involve restoring an old canal. Training workshops make things for sale, so you learn how to use tools, lathes and the like. One training workshop restores old railway engines. There is also a scheme in some areas

called *Service Away From Home* which is for young people who want to do community service away from the area they live in.

- *Work Introduction Courses*. These are short courses, which are useful if you are not sure what kind of work you want to do. They give you a chance to try your hand at several different kinds of skill; they offer help with reading, writing and numbers too, if you want it.

General information

- The government has said that every young person who is still unemployed by Christmas after leaving school will be guaranteed a place on YOP. Altogether, you can spend up to *12 months* in YOP (and sometimes even longer, if you still have difficulties in finding work after leaving a scheme). So you can do a short training course, and then go on to do work experience afterwards, if you still can't find a job.

- On all the schemes, if you find a job during the scheme you can leave at once. Every scheme gives you a certificate when you leave which describes the kind of work you have been doing. You can show this to an employer when you're looking for work.

- Only work experience and the short training courses exist in every area; if there is a short course which you want to go to which is too far to travel to every day, you can receive a special allowance so that you can live away from home — and you'll get help finding somewhere.

- The place to go to find out about YOP is the Careers Service, who have all the details. Phone up, make an appointment, then go in and have a proper talk. If you want to do work experience, think about what kind of work you'd like to get experience of before you go, as this will help a lot.

Community industry

This is very similar to some YOP schemes, although it is organised separately. When you are in the Careers Office, ask if there are any Community Industry schemes locally. You might learn basic building and decorating skills, by renovating an old church hall, or outdoor work skills, by landscaping the grounds of an old people's home.

3. GET A PLACE AT COLLEGE

The full details about this possibility are in Chapter 7. Remember; college is not school. Far from it. It is a far more adult place. If you aren't sure if you want to study or not, don't just think, 'Oh, I'm not interested.' Go and arrange a visit, and meet someone who can tell you what kind of courses you could do.

Once you are over 19, the courses will start costing you money. Before you are 19, they are *free* — so make the most of your chance.

4. IF YOU ARE COMING UP TO 19, GET A PLACE TO TRAIN ON A TOPS COURSE

The full details about TOPS courses are in Chapter 7. Quite a number of TOPS courses have a waiting list, even once you are accepted. Ask the Careers Office (or if they're not sure, try the Jobcentre), and then if you have a few months to fill in, do something out of the 'active unemployment' notes below.

5. ACTIVE UNEMPLOYMENT – DO IT YOURSELF

'We got together and advertised ourselves "for hire" at £1 an hour. We all earn the £4 that is allowed now, which pays for other activities. All sorts of work – gardening, painting, shifting rubble, babysitting – things like that.' Rob, 18.

'I use the "21 hours a week rule" to go to college, where I'm taking two 'O' levels, and learning to weld.' Valerie, 17.

'We've formed a local group to campaign about unemployment. We're demanding free entry to the swimming pool, half-price football matches, and free bus passes to help us look for work. The papers gave us quite a good write-up.' Dave, Josie, Lyn, Mitch and Kevin.

'I went away on a workcamp for three weeks in the summer. It was fantastic. We were up on Snowdon, in Wales, doing work to look after the place.' Pete, 20.

'We've formed our own motor-bike repair workshop; I've got this mate who really knows his stuff, and he comes and teaches us two nights a week.' Stu, Bill, Rich.

'We all go rock-climbing every weekend with the local club. We're thinking of entering the Duke of Edinburgh's award. It would be something to aim for.' Valerie, Rosie, Sonia.

'I got in touch with the Trades Council and found out that they are trying to set up a centre for unemployed people, to give advice, organise protests, fight prejudice and things like that. I help out there now, which is good, and feels right. I've learned how to use a duplicator, and I've started giving advice to other people about their benefits problems.' Danny, 19.

'I'm with a group that's making plans to start up a co-operative business making baby-carriers. We're getting lots of advice from people who know what they're talking about.' Carmen, 18.

Those are some of the ideas – there could be 100 more.

General information

- *Earning money*. Anyone on Supplementary Benefit can earn up to £4 a week after taking away all expenses associated with work, before benefit is reduced. If you wanted to, you could earn more than that, and ask for the extra 'pay' to be donated to a local charity, to the youth club, or to some other group. Remember to deduct any advertising, travel, telephone or special clothing expenses from your earnings, so that you end up with a clear £4 in your pocket. If a group were to set up a small 'agency', as young people in Liverpool have done, each person might earn £5, and pay £1 a week into the agency for the expenses of running it.

- *21 hours a week studying.* Go to the college and see what courses they offer. You can do 21 hours a week without losing your benefit, but not all authorities know this, so you may have to appeal for your rights.

- *Forming a group to campaign against unemployment.* You'll find ideas and advice in Chapter 12. Campaigning for cheaper prices, helping each other with benefits, organising protest marches, persuading local councillors and the MP to do something, making sure the papers print fair stories, working with local trade unions to develop joint policies to fight unemployment and help the unemployed.

- *Joining a workcamp.* On a workcamp a number of people, mostly young, work together for two to three weeks on a scheme of some kind or other. They sometimes take place in cities, sometimes way out in the remote country. There's no pay, but all your costs are met while you are there. For details of camps, write to: International Service Volunteers (53 Regent Street, Leicester LE1 6YL), or Friends Service Council (Euston Road, London NW1 2BJ), or Interaction (Talacre Open Space, 15 Wilkin Street, London NW5 3NG).

- *Other kinds of voluntary work.* There really are lots of different kinds of work to be had, such as the Sue Ryder homes, which need volunteers to help care for disabled ex-service people, or the Edinburgh Cyrenians, who need volunteer workers for their city house and farm community.

  To find out all the details of this kind of voluntary job, send a large stamped addressed envelope to the Voluntary Service Opportunity Register (NYB, 17 Albion Street, Leicester) and they will send you a *free* booklet, which they publish every spring.

- *Motor-bike repair workshop.* If there is a small group of people wanting such a workshop, you could go about getting help like this:
  - Ask the adult education people (Chapter 7 — address p. vii) to put on an evening class for you.
  - Try the police — they might release someone to work with you.
  - Ask the local motor-bike shop if one of their mechanics would help you.
  - Ask one of the youth club workers if he would help you get a group going.

    Allow time for all these suggestions — it might take a month or so to find a place, and get it organised.

- *Centre for unemployed people.* Write to the Newcastle Centre (5 Queen Street, Newcastle on Tyne) for details about their centre, and then contact the secretary of your own local Trades Council. Tell him about the centre, and say you would like them to organise something similar. Trades Councils are associations of local trade unions. The TUC (Congress House, Gt Russell Street, London WC1) can send you details of your local Trades Council.

- *Planning to set up a co-operative.* You can find the addresses of people who could help you at the end of Chapter 9.

- *Summer jobs*. Ask your library to get copies of *Working Holidays* (85p from CBEVE, 43 Dorset Street, London W1H 3EN), *Time Between — A Guide for Work and Service* (95p from CRAC, Bateman Street, Cambridge CB2 1LZ) and *Summer Jobs in Britain* (£2.25 from Vacation Work, 9 Park End Street, Oxford OX1 1HJ). These give details and addresses of paid work on farms, fruit-picking, in hotels, domestic work, workcamps, voluntary work openings, adventure courses and holidays, overseas work and workcamps, au pair work, etc. They will completely change your ideas about what is possible.

- *Community Service Volunteers*. CSV finds voluntary work positions for 2000 people a year. Volunteers receive a small weekly allowance plus full board and lodging and travel expenses. They never turn anyone away who applies. You can choose what you want to do from a great variety of work. (CSV, 237 Pentonville Road, London N1 9JN.)

- *What about my signing on, if I go away on one of these things?* If you will be earning money, then you should sign off the dole and sign on again when you return. Explain what you are doing to the UBO.

   If you will not be earning any money (e.g. going on a workcamp) and you will be away for less than four weeks, ask the UBO for a *Holiday Form*. Fill it in, leaving an address where they can find you, in case a job turns up for you. (!) Then you can pick up your benefit when you get back. So long as you can show them that you are 'available for work' and able to return at once if need be, you should have no problems.

- *Travel. Hitch-hiker's Manual*, by Simon Calder (£1.75 from Vacation Work, 9 Park End Street, Oxford). Long distance coach; advertise for a lift on a local notice-board.

- *Local activities.* Chapter 8 is full of ideas. Go through it, and find out what interests you. Then use Chapter 2, as well.

Other kinds of work

- *Building sites* take on labourers at 8.00 am in the morning. *Hotels and cafes* take on extra staff in the summer. *Farms* take on extra labour at harvest time; potato-picking in Pembrokeshire, Cornwall and on the East coast in early June, elsewhere inland later; soft fruit picking, early July; plums and apples later in Worcestershire. This kind of work is all piece-work, so it takes a week or two before you are quick enough to earn any big money.

- *Residential jobs. The Lady* carries small ads for living-in jobs as maids, au-pairs, home-helps, etc. Social Service Departments can give you lists of private homes for old people and children where there might be work. *Horse and Hound* has small ads for living-in jobs with horses and other out-door work. You could advertise in either of these magazines yourself e.g.:

```
┌─────────────────────────────────────┐
│        Strong lad / healthy lass     │
│              seeks work              │
│                                      │
│      Will consider anything legal.   │
│                     Box 300.         │
└─────────────────────────────────────┘
```

- *Au Pair work.* For jobs in Europe (if you can understand enough of the language of the country) write to the International Catholic Society for Girls (Room 17, 1st floor, 39 Victoria Street, London SW1H 0FE) with a stamped addressed envelope. These are three-month posts in France, Belgium and Italy and six-month posts in these countries plus Austria, Germany, Spain, Switzerland.

6. FORM A SELF-HELP GROUP WITH OTHER YOUNG UNEMPLOYED PEOPLE

If you can do this, then life might become easier for you. Chapter 5 has got full advice and suggestions about who might help you get such a group going. The group could help you to find out about college, apply for jobs, etc. and maybe become a wider 'action' group as well, to try to do something about local unemployment. It's only when people start protesting about unemployment that anything will really start to happen.

- Write and tell us about anything interesting you get up to, while you are unemployed, and we might be able to tell other people about it, perhaps through one of the national daily papers.

Chapter 6B: Notes for unemployed women

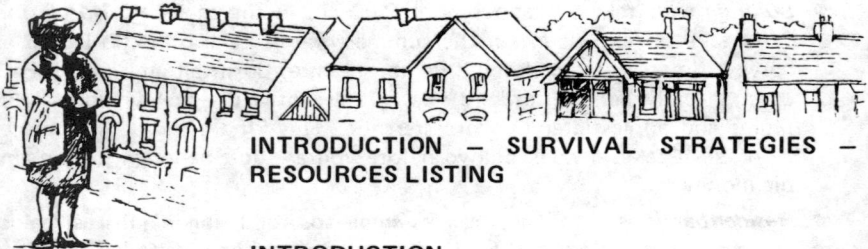

INTRODUCTION — SURVIVAL STRATEGIES — RESOURCES LISTING

INTRODUCTION

As unemployment rises, so do the little voices that say 'the women should go back home, and leave the jobs for the men'. In a number of places, the women are the first to be told they are no longer needed, and you can almost *hear* the employer saying to himself 'The job isn't so important to them, so we'll throw them out first.'

If some men feel so strongly that children need a parent at home, they are at liberty to stay at home themselves. It is very important that the unemployment crisis is not used to erode women's rights, and their freedom to choose what they do with their lives.

Women's unemployment is rising faster than men's. The jobs that they get are lower paid, and offer much less opportunity for advance. Much is talked about equality, but when it comes down to it, women so often still end up in traditional women's jobs as clerks, cashiers, secretaries, maids, cleaners, shop assistants, nurses and factory hands.

The microprocessor revolution is going to cut quite sharply into one particular area that has traditionally been 'women's' — typing — which could lose up to 40% of its workforce to the word processor. Much of the low-paid routine factory work is also in line for takeover by automation and robots.

Nothing good is going to happen without a struggle. History hasn't changed. Only the nature of the struggle changes, as the times change.

SURVIVAL STRATEGIES

In a realistic survival strategy for the 1980s, these might be the choices:

1. Planning on getting a *full-time job*, in competition with the men. This means being sharper, brighter and better than the men. If women give up their assaults on men's jobs then woe betide us all, because those jobs *need* women to prise apart the self-satisfied male 'we-know-it-all' attitudes that still prevail. When our organisations and enterprises begin to be a real partnership between men and women, then we'll be getting somewhere.

   Chapter 4 has got advice on practical jobhunting, and see also Resources A, B, C, D, E, F, G, H, I.

2. Aiming to get decent *part-time work*, and if this isn't going to mean the normal choice of dead-end jobs, it will have to mean that *jobsharing* becomes more and more a part of everyday life, so that two people can share a good job together, and work out their hours or weeks as they find best. See Chapter 11.

   There is also *home-work*, which has traditionally been one of the most exploited areas of women's work. See Resource L.

3. Becoming *self-employed*, setting up a new *co-operative*, as 'Little Women' did, or converting an existing way of working into a co-operative, as the Kennington Office Cleaners Co-operative did. It would be interesting to see a breakdown of the facts about new small businesses, because the hunch has been expressed that the success rate among women is higher than it is among men, for reasons that have yet to be researched. If setting up new enterprises or

becoming self-employed is to become a regular part of women's life, the special self-help groups mentioned in Chapter 9 could play an important role. General advice in Chapter 9.

4. *Involvement in local neighbourhood action and activities,* so that if there is simply no work going, life need still not return to being a trap. There is plenty of work waiting to be done, in the way (e.g.) that the people of Craigmillar, in Edinburgh, have done. This work can be involvement in play activities, tenants associations, setting up community enterprises so that the economic basis of a locality is rebuilt from the roots up. See Chapter 8 for ideas and contact addresses, and Resource R, S, T.

5. *Involvement in women's groups,* that help women to build up their self-confidence, find their voices, and discover their strength. In a male-dominated society, these groups are vital for developing local strategies for action and activity.

There are women's groups in most cities and towns, and in the larger cities there are often women's centres too. Such groups would also help the kind of support group suggested in Chapter 5.

If you do not know of any local group, write to the Spare Rib Collective enclosing a stamped addressed envelope and asking for details of the nearest group (Resource N).

6. *Study, education, training.* There are a number of special courses for women — the TOPS 'Wider Opportunities for Women' courses, and the 'Return to Work' or 'Return to Study' courses. There are ways of studying at home too — and local self-help study circles can always be set up, if there is nothing else available. See Resources I, J, K.

Making choices

If you are trying to make choices about what you want to do next, you might find it helpful to consider the six options outlined above (not all mutually exclusive, obviously), and place them in your own order of importance from 1—6. You might then find it interesting to compare and discuss your notes with friends who may be unemployed.

RESOURCES LISTING

A.   The Equal Opportunities Commission has produced a *free* booklet called *Fresh Start*, which covers most things to do with training and education for women who are returning to work, or who are thinking about a change of work. It is available from Jobcentres, or from the Equal Opportunities Commission, Overseas House, Quay Street, Manchester M3 3HN. *Recommended.

B.   *Women at Work: A Handbook for the Working Woman*, by Jenny Glew (Pitman, 95p) covers many useful things not dealt with in this handbook.

C. If you are considering working in a traditionally 'male' job, try *Equal at Work: Women in Men's Jobs*, by Anna Coote (Collins, £1.50). She writes about individual women working as an engineer, lorry driver, solicitor, furniture restorer, forester, plumber, airline pilot, sales rep, astronomer, car mechanic, accountant, craftsman and gardener. The book also covers education, training and equal rights.

D. There is a support network called *Women and Manual Trades*, which publishes a regular newsletter, organises conferences, etc. They keep a 'Register of Women in Manual Trades', to put women in similar areas or occupations in touch with each other. Contact them: c/o 1st of May Bookshop (43 Candlemaker Row, Edinburgh), or c/o 58 Bexley Avenue, Leeds 8, or c/o 40 Dale Street, London W4.

E. *Equal Opportunities: A Careers Guide for Women and Men*, by Ruth Miller (Penguin, £1.95). Thorough and invaluable. Also useful advice concerning part-time work, and for women who are returning to work after a break.

F. *Happy Return*, by Margaret Korving (BBC Publications, £1.25), who produces careers programmes for the BBC. For women returning to work.

G. *Women x Two: How to Cope with a Double Life*, by Mary Kenny (Hamlyn, 90p). Useful reading for anyone wanting to combine work and children.

H. Women in London with a little cash to spare can seek careers advice from the *National Advisory Centre on Careers for Women* (251 Brompton Road, London SW3 Tel: 01-589 9237). This is a long-established charity which offers good professional advice.

I. *Wider Opportunities for Women* courses (WOW) are run by TOPS, the Government training organisation. (See Chapter 7.) In some areas they are called 'New Opportunities for Women', or something similar. If you can get a place on one of these courses (and there are not very many of them yet), you will be paid an allowance at the TOPS rate. (Leaflet L 91 from Jobcentres gives the details.)

The courses last for up to six weeks (or up to 12 weeks part-time in some areas, for women unable to take a full-time course). They give you a chance to try your hand at a wide range of different kinds of work — welding, car maintenance, TV repair, hairdressing, typing, office work, catering, work in a design office, carpentry, gardening, hotel work, etc. You will be offered relevant information about jobs, training, education, etc. and you will be able to visit factories, offices, day nurseries, etc. You will also get advice concerning children, tax, application forms and interviews, effective speaking, etc.

You can apply for a WOW course if:

- You have not worked for at least two years or have never had a job because of your domestic responsibilities.

- You are available for work.
- You are at least 19 years old and have been away from full-time education for more than two years.
- You don't know or are not sure what sort of work you'd like to do.

'Open Forum' groups: all the women on a course are also given the chance to join an 'Open Forum' group, where they get together to talk things over, and to share their feelings, ideas, problems, etc. (e.g. how to cope with picking the children up from school). The groups are very helpful for discovering that you are not alone in feeling guilty, inadequate or nervous about returning to work.

How to find out more: ask at the Jobcentre. Some women arrange to take week-time lodgings for the duration of the course, if the nearest course is some distance from home.

J.   *'Return to Work' and 'Return to Study' courses*
'There is a vast variety of courses for people who left school some years ago, perhaps with no or only a few 'O' levels, or who have a rusty 'A' level or two, and who now want to train for a new career and/or go on to higher education, or just to keep their brains in trim.

'Courses very enormously in content and organisation, but they all have one thing in common: they act as confidence restorers.' (From *Second Chance Education*, by Ruth Miller. See Chapter 7.)

The courses are not always well advertised, so you should ask at your local college. They are often considered as 'fringe activities', and so are not well known. The courses have no entry qualifications, and classes are small.

K.   If you would like to return to study, but are unable to get to a college easily, you may be able to *study at home*. See the 'Study at home' section of Chapter 7.

L.   *The Home-Workers Association* (9 Poland Street, London W1V 3DG) exists to support people who do piece-work at home (often at very low rates). They publish a newsletter, called *Home News*. The group is campaigning for higher wages and for better organisation among home-workers. In order to be an organisation with any influence, it needs home-workers to get in touch and help the campaign.

M.   *Earning Money at Home* (£2.95, Consumers' Association) gives ideas and advice that might be useful. Available from bookshops and libraries.

N.   The magazine *Spare Rib* contains a lot of material of interest to women, including features on employment and women's projects, and regular listings of meetings, events, group addresses, practical advice, etc. It is available monthly from newsagents, price 40p, or £6.50 for a year's postal subscription from 27 Clerkenwell Close, London EC1R 0AT.

0.   The magazine *Women's Voice* is more openly political. It has a strong fighting spirit, and runs regular features on unemployment, women's rights, etc. Available from some newsagents, or £3.60 for 12 issues from Women's Voice, PO Box 82, London E2.

P.   *Job Massacre at the Office* is 'an informative, easy to read pamphlet that no office worker can do without. Facts and figures on word processors and on how to deal with them' including detailed trade union policy and negotiation positions, and advice on how to organise at the workplace. Price 40p including post from Word Processor Pamphlet, Women's Voice, Box 82, London E2.

Q.   *The Directory of Social Change: Women* (1978, £3.95) is a complete encyclopedia of useful information, ideas and addresses, covering work, rights, health, relationships, life-styles, therapy, child care, etc. Available from bookshops, or from the Directory of Social Change, 9 Mansfield Place, London NW3.

R.   *The Women and Employment Project* (74 Deptford High Street, London SE8) works to investigate and improve employment and training opportunities open to women in Lewisham. They have published a local 'alternative prospectus' for women, including information on ways by which women can get into engineering, electronics and manual trades; also a handbook of child-care facilities. Funded by the Docklands Urban Aid Programme, and set up by a group of local women.

S.   *Milton Keynes Women and Work* (7 Wetherburn Court, Bletchley, Milton Keynes) has developed a resources and information base for local people to use; runs a drop-in advice centre; works to get more training facilities for women. Funded by the MSC through STEP.

T.   *Lambeth Women's Workshop* (C22, Park Hall Trading Estate, Martell Hall, London SE21 8EA). Set up by women from Lambeth Women's Aid Group in response to the special problems of women running one-parent families. Offers free training in carpentry and joinery skills. Funded under the Inner City Programme.

There are plans to create an ambitious women's training workshop in East Leeds. Contact the workshop c/o 120 Markham Avenue, Leeds 8.

(Please send stamped addressed envelopes to all projects, when writing for information.)

See also Chapter 4 for information on sex discrimination.

Chapter 6C: Notes for disabled people

SPECIAL HELP — SOURCES OF ADVICE AND INFORMATION

Needless to say, circumstances are always that much harder for people with a disability than for those without.

There are various ways in which disabled people can get special help in finding suitable employment.

1. The Employment Services Division employs *Disablement Resettlement Officers* (DROs) whose job it is to assess, advise and help any person who is disabled. They can be found through the Jobcentre. The ESD also runs Employment Rehabilitation Centres, where they put on special residential and non-residential training courses for disabled people. 'Disablement' can cover such things as recurring depression or nervousness as well as physical disability. If a person has a problem which effectively prevents him or her from settling into steady employment, help from the DRO may be possible.

2. If a person who is disabled carries a *Green Disablement Card* which certifies that the disability is officially recognised, employers can then claim the person as one of their required quota of disabled people. Also, if a disability is obvious, an employer may want to see the green card, and without it the person may not get the job.

3. Disabled people over 16 can join any *TOPS course* without waiting until they are 19. The other rules about TOPS courses are also suspended — you can take longer over a course and receive extra help with housing and transport problems, for instance. See the leaflet *Training Opportunities for Disabled People*, available at Jobcentres.

4. The leaflet HB1 *Help for Handicapped people* (from post offices) gives details of what is available through the DHSS by way of extra benefits and services. The Social Services Department will probably also be able to help. See also the leaflet HB2 *Aids for the Disabled*.

5. *Job Introduction for Disabled Workers* is a scheme that encourages employers to take disabled people on trial, in the hope that this might lead to permanent

employment. The employer receives £30 a week for 6 weeks. Details from Jobcentres.

6. Under certain circumstances a registered disabled person who is unable to use public transport can get *help with travel costs* to and from work. See leaflet DPL 13, from Jobcentres.

7. *Employers can receive a wide range of aids* from the Manpower Services Commission to help a disabled person more easily — special fittings for tools and typewriters, special machine modifications, etc. Many employers may not know this, and may therefore think that it would be very difficult to take on a disabled person. If you carry leaflet EPL 71 (from Jobcentres) with you when you go jobhunting, this might help. Get advice from the DROs, too. Adaptations to buildings necessary for the disabled employee may also be carried out by MSC.

SOURCES OF ADVICE AND INFORMATION

The Source Book for the Disabled, edited by Glorya Hale (Paddington Press, 1979, £4.95). A general resource for the physically disabled. Covers such subjects as employment, leisure and sexuality, and is particularly good on the subject of aids and adaptations to the home.

Disability Rights Handbook (Disability Alliance, 1979, £1, from 1 Cambridge Terrace, London NW1 4JL, or £3 for the Handbook plus updating bulletins for a year). Reliable, accurate, and thorough.

Rehabilitation, Retraining, Resettlement, published by the MSC, gives vital information and advice on training and employment. Ask at Jobcentres. MSC also publishes leaflets on *Employing Disabled People* (EPL 61 — mention it to an employer), *Into Work* (EPL 41) as well as *Aids and Adaptations* (EPL 71).

A Handbook for Parents With a Handicapped Child, by Judith Stone and Felicity Taylor (Arrow Books, £2.50). Full of useful information for the handicapped adult as well, and not just for parents. Useful information on careers guidance, education, training, employment, leisure and holidays.

The Disabled Living Foundation, 346 Kensington High Street, London W14 8NS offers an information service, covering almost every subject with guides and information sheets, including one on *Further, Adult and Higher Education — Assessment and Training for the Physically Handicapped* which is *essential reading*. Members of the general public may use the service free of charge.

Employment Services for the Disadvantaged (£1, from Personal Social Services Council, 2-16 Torrington Place, London WC1E 7HN). This report looks in detail at the various problems and services which exist for the disadvantaged.

Disabilities Unlimited is a catalyst organisation that hopes to show that by working together, with commitment and determination, it is possible to become independent, and to lead a rewarding life. They are keen to contact other disabled and able-bodied people who will join with them, to form residential and working collectives. Send a stamped addressed envelope for further information, or phone to find out more, to Ian King, Disabilities Unlimited (22 Dane Road, Margate, Kent: 0843-25902).

Alastair Kent, at the Cambridgeshire Careers Centre (County Hall, Hobson Street, Cambridge) is compiling a *Handbook of Initiatives* through which disabled people are tackling the problems of unemployment. Contact him if you want to discuss any ideas, or tell him what you are doing.

Chapter 6D: Early retirement: Notes for older people

SOURCES OF ADVICE AND INFORMATION

Retirement is a strange idea. It never existed in societies until very recent times. Why should people suddenly stop working, just because they have passed a certain age? Why should they or anyone else who is without a normal paid 'job' be made to feel that they are outside normal society?

The pressures for early retirement are likely to increase so here are details of a few schemes which offer alternatives:

- *Retired Executives Action Clearing House (REACH)* links retired professionals with charities, voluntary organisations and community groups which need their skills, but cannot afford to pay for them. Many early-retired people work for such organisations for expenses only. (REACH, 93 Highgate West Hill, London N6 6EH.)
- *Age Concern* has a number of enterprising self-help schemes to enable the early-retired and the retired to use their skills and abilities actively in the community. (Age Concern, Bernard Sunley House, 60 Pitcairn Road, Mitcham, Surrey.)
- *Success After Sixty* is a special agency for jobs for the retired and the early-retired. (Success after Sixty, 40/41 Old Bond Street, London W1X 3AF.)
- *The Time of Your Life* is a handbook for retirement, including practical sections on finance, health, activities, hobbies and profitable employment. (£2.25 from Help the Aged, 218 Upper Street, London W1.)
- *Trades Skills Register:* the Estate Department at Dartington Hall in South Devon has established a service for people living on pensions who can't afford the cost of hiring tradesmen to carry out maintenance and repair work on their homes. It consists simply of a register of retired people who are willing to put their skills at the service of other retired people. How much they choose to charge is up to them.

  Anyone with initiative could launch a similar scheme: all that it requires is some publicity, some leafleting and door-knocking, and keeping a register in a place where people could see it — or duplicating copies, so that

every member of the scheme could have one. If members were still drawing Unemployment Benefit or Supplementary Benefit, they would need to
keep any earnings within the legal limits (see Chapter 10), unless and until
that system is changed.

- *Rebuilding supports:* if you have little prospect of finding another job at
all before you reach retirement age, the advice in Chapter 2 of this Handbook becomes very important. The transition from work to non-work can
be very destructive if the supports outlined in that chapter are not rebuilt.
If a positive attitude is taken this crisis can be turned into an opportunity,
and the transition to a new pattern of living and working can be achieved
successfully.
- Write to the *Pre-Retirement Association of Great Britain and N. Ireland*,
19, Undine Street, Tooting, London SW17 8PP, for a list of local groups,
and details of activities.

Chapter 6E: Notes for redundant workers

BEFORE THE REDUNDANCIES HAPPEN — PER—
SONAL STRATEGIES — INFORMATION

BEFORE THE REDUNDANCIES HAPPEN

A thorough anti-redundancy strategy might involve consideration of the following points, *in addition* to the normal strategies in existing use:

- The trade union branch (or a workers/staff council) must try to get access
to all the facts, in order to be able to develop appropriate policies. If you
don't have information, you will be powerless. A trade union research department may be able to advise. Is the firm really in severe difficulties?
Are the difficulties being caused by internal mismanagement? Is one
branch being run down?
- The entire workforce ought to be consulted about ways in which efficiency
can be improved, new ideas brought in, or how the firm could diversify
into new, more profitable areas. Workers on the shop-floor often have
good ideas for change which never get through to management, for various
reasons. When the shop stewards combine committee at Lucas Aerospace
were facing redundancies several years ago, they circulated the entire
workforce for their ideas about socially useful products that the firm
could make, and received suggestions back that enabled them to put together an alternative 'Corporate Plan', consisting of 200 possible products.
If the overall problem is presented as a *collective* problem, then the stage

is open for the development of collective and constructive solutions.
- Once the problem is being looked at collectively, an option such as the entire workforce taking a 10% cut in wages, instead of a 10% cut in numbers, can be considered. One textile firm in Yorkshire has done just this, since the workers were all so keen to keep the firm going.
- Similarly, some workers might be interested in jobsharing, by which two workers agree to share one job between them. It is often the case that some workers would prefer this way of working, anyway, since it leaves them half the week (or every other week) for other kinds of work, or for self-employment, etc. See Chapter 4 for advice.
- *All* firms (especially the larger ones) ought to be working out ideas and plans *now* for future cut-backs in order to widen their range of options when difficulties arise. Trade unions could be pressing for firms to come up with positive and creative approaches. Some firms, for instance, have started encouraging employees to think about ways in which the company might foster new small enterprises under a protective family umbrella which could later become autonomous. ICI are doing some creative work along these lines.
- When redundancies are inevitable, trade unions might consider widening the normal basis of redundancy negotiations, and ask the firm to include two 'Social Clauses':
  (a)   Personal career and jobhunting counselling as well as money for redundant workers. Professionally-led workshops can help people approach and use redundancy positively, instead of with despair. Some of the methods in this Handbook can be adapted quite easily to workshop-use.
  (b)   Funds for the development of new community-based initiatives, to help the community develop its own local economic strength. Chapter 12 mentions some of the work already proceeding in this field; this urgently needs a more secure funding basis.
       In general, there is an urgent need for constructive thinking and positive responses.

PERSONAL STRATEGIES

Once the redundancy has happened, what can you do? It may take quite a while to re-establish yourself in a new job. This is not being pessimistic — just realistic. The times are not easy. However, this being so, the thing to do is to ask yourself, 'How can I therefore use this period of time to my advantage?' If used constructively, the six months or longer during which you are without a normal job could become a very important and creative period of your life.
- *Rebuild your supports*. Chapter 2 runs through the various ways in which people who lose their jobs also lose important personal supports, and shows how you can rebuild these supports for yourself, so that your life can remain satisfying and fulfilling. Chapter 8 suggests various ways in which skills can be maintained and developed, new contacts made, and ful-

filment found through challenging work on an unpaid basis in the local community.

- *Self-organisation*. This is very important. Without it, things can easily start to slip downhill. Chapter 2 can help.
- *Join or form a support group*. This will help with self-organisation, and may well lead to opportunities and activities that would otherwise have remained undeveloped. Chapter 5 gives advice.
- *Become clear about your own personal choices and directions*. It is always easier to go somewhere if you know where you want to go. Use the 'Difficult choices' notes and Chapter 3 to help. Another useful suggestion is that you take a sheet of paper, and simply ask yourself, 'What are my three main priorities in life at the moment?' Take some time to think about this, and then write down what they are. Then ask yourself, 'What am I doing to fulfil them?' and again write down your decisions. This kind of activity is useful to do with a group, if this is possible.
- Do not forget the possibilities of *jobsharing* (see Chapter 4), of *self-employment* (see Chapter 9), and of *part-time work* (Chapter 10 explains under what conditions you can take *part-time work*).
- *Your skills* could be very useful to local groups in the community, which are always in need of people with practical skills. You could help them, and keep your own hand in at the same time. By doing this, you are not only helping them and helping yourself; you are also increasing your chances of finding a new job, simply by keeping active, by making new contacts, and by gaining new areas of experience. Some firms deliberately look out for people who have broadened their experience in this way.
- *Local activities against unemployment*. Chapter 12 has some ideas around which local campaigns can develop. It is the people who are themselves suffering from being out of work who have perhaps the most to give to a determined effort to get major changes made.
- *Creative career development (and life development)*. OK, it's hard to see how time out of work can be positive. But it could — and *must* — it will become a time for exploring new avenues, for developing new experiences, and for building a more solid personal basis for life. We all tend to be very dependent on our jobs to give us a sense of who we are, and what we are doing in life. It can actually be very good for us to have to face up to this, through the job being taken away for a period of time, so that we remember that we have to have our own independent lives, and not rely so much on a job to bring us meaning, purpose and identity in life.

INFORMATION

Books
Focus on Redundancy (Kogan Page)

The Consumers' Association publish a small book called *Dismissal, Redundancy and Job Hunting* (price £1.50), which explains legal aspects concerning redundancy and dismissal which this Handbook does not go into. (Available through libraries, bookshops, or the Consumers' Association, Caxton Hill, Hertford SG13 7LZ.)

The Department of Employment publishes a series of booklets about the Employment Protection (Consolidation) Act, 1978, which you can obtain free from the Jobcentre. In particular:

No. 2: *Procedure for handling redundancies.*

No. 6: *Facing Redundancy? Time off for job hunting or to arrange training.*
        The Redundancy Payments Scheme (complete details + basis of payments).
        Dismissal — Employees' Rights (complete details).

Redundancy payments

The minimum qualifying period of employment for redundancy payment is two years — 104 weeks continuous service. Part-time workers must have worked for 16 hours a week or more (or eight hours per week for five years or more).

If an employer offers suitable alternative employment, which is unreasonably refused, payment cannot be claimed. Employees have a four-week statutory trial period in which to find out if the job fulfils its promise or not. If it does not, you may leave and still claim redundancy payment.

There are various ways of interpreting these and other provisions, and if you are in any uncertainty, you should consult one of the booklets mentioned above.

Tax and benefits

Any redundancy payment which you receive is free of tax.

If you are entitled to redundancy payment, you are also entitled to receive Unemployment Benefit.

You are entitled to claim a rent and rate rebate while on Unemployment Benefit.

You are also allowed to work part-time while on Unemployment Benefit, if you forgo benefit for the days on which you work. (You will need to calculate whether the pay from your part-time job will come to less or more than one-sixth of the weekly benefit.)

There are three recent changes in 'the system' which you might not fully appreciate, which are fairly serious:

1. Earnings-related supplement is being phased out in 1981, and will be abolished in April 1982. Thereafter, anyone becoming unemployed will go straight onto the basic level of Unemployment Benefit (if conditions are satisfied — see Chapter 10).

2. The basic level of Unemployment Benefit is deliberately being reduced by five percentage points less than inflation each year, in order that the unemployed should 'shoulder their share of the cuts' (as if being unemployed was not a sufficient share already). This may mean that more people on Unemployment Benefit will need to resort to Supplementary Benefit in order to have suffi-

cient day-to-day income — if they qualify. Unemployment Benefit runs out after a year anyway, after which people normally apply for Supplementary Benefit.

3. Anyone who has over £2,000 in savings is now *completely disqualified* from receiving any Supplementary Benefit. Your house and car, etc. are not counted as savings, but a life assurance policy is, which is particularly cruel and illogical.

It is not difficult to see what this means. It means that many people will get no benefit at all after their first year of unemployment, until their savings have fallen to the last £2,000. They will, however, still be eligible for rent and rate rebates, and child benefit. They will also be free to take any kind of part-time work, since they will no longer be constrained by any government benefits system. House owners receiving no benefit ought perhaps to negotiate with their building societies to see if they could shift over to the 'Option' mortgage scheme, which is designed for non-tax-payers.

Full Benefits details — Chapter 10.

Chapter 6F: Notes for professional and executive workers

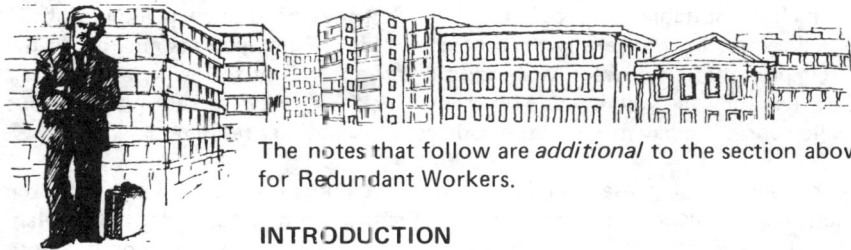

The notes that follow are *additional* to the section above for Redundant Workers.

INTRODUCTION

A major article in a Sunday paper n 1980 was headed 'The Loneliness of the Redundant Executive'. The writer, Robert Chesshyre, wrote it after an unemployed executive living in an affluent suburb rang 'to make a plea for some reporting about those in his own lonely and bewildering predicament'.

Being unemployed is just as hard for a manager whose firm has gone to the wall, or who has been dismissed, as it is for the redundant steel-worker. Many professional people who lose their jobs find that one of the hardest things is the feeling of isolation. The contrast between then and now is acute. All the perks, the status, have gone, and the outlook for finding a new job can sometimes seem pretty bleak.

The overall philosophy that underlies this Handbook still applies as much as ever, however. Focus on the negative, and you will feel negative, and you will actually lose energy. Focus on the positive, seek the ways in which being unemployed brings you opportunities, and you will find that things are not so bad, and that you have got the personal resources to pull through.

PROFESSIONAL AND EXECUTIVE RECRUITMENT

PER is a branch of the Employment Services Division of the MSC which offers a specialist job-finding service to professional and executive people. Anyone who is registered with PER benefits from a lot of help that is simply not available to other unemployed people:

1. PER's free introductory *folder of information*, which includes a copy of a *Guide to Successful Jobhunting*.

2. A free weekly jobs magazine that you receive through the post, called *Executive Post*, which details all the jobs on PER's lists in either (a) sales, marketing, purchasing and distribution, or (b) managerial, executive, professional, technical and scientific occupations. It also carries articles on jobhunting, etc.

3. Free half-day 'Job Hunting Seminars', which include an opportunity to get help in preparing a plan of action for your own job search.

4. Three-day 'Self Presentation Courses', available to candidates who are unemployed or facing redundancy and who have a clear idea of what job they want but who need to improve their jobhunting skills.

5. Special courses, lasting two weeks, called 'Career Development Courses', which are suitable for people who feel that they need to change their jobs in mid-career, or who are in need of counselling. These courses enlarge on the self presentation courses, with additional opportunity to review one's career so far and one's abilities, with a view to taking new directions. Generous allowances are payable, and the courses are sometimes residential.

The value of all these courses should not be under-estimated. Apart from the practical skills which you can learn and the opportunity for constructive planning, the opportunity to tackle your own problems in companionship with others in similar situations is valuable. In some areas of the country these courses are often under-subscribed.

PER's services are available to anyone who is seeking a job who has 'A' levels or above, or the equivalent. Interviews with PER consultants can be arranged by appointment. You can contact PER through the Jobcentre, or through the phone-book.

OTHER INFORMATION

- *TOPS higher courses*. You can obtain a list of TOPS higher level courses either through PER or through the TOPS District Office. Examples of the kind of higher level course available are management studies, computer programming, higher certificate in electronics, small business management. See Chapter 7 for full TOPS details.

● *Career consultancy firms.* There is a growing number of private firms specialising in career counselling, or consultancy, all of which have something of value to offer, at varying costs.

CEPEC, the 'Centre for Professional and Executive Career Development and Counselling', for example, offers employers and individual men and women a range of services to assist them in career development and in a career or employment crisis. It is concerned with career reviews, mid-life changes, career planning and redundancy. Through personal counselling interviews, CEPEC staff are able (a) to help people in an employment crisis to identify any personal factors that may be playing a hidden role in the client's unemployment, and (b) to offer help in such areas as employment market information, employment opportunities, training opportunities, personal projection and finance planning. You can write for further information about CEPEC to The Director, CEPEC, Sundridge Park, Bromley, Kent BR1 3JW.

The Vocational Guidance Association (7 Harley House, Upper Harley Street, London NW1) will take you through psychological tests and draw up personality, interest and aptitude charts, on the basis of which they will then suggest a suitable line of work.

In most larger urban areas you will find details of Career Consultancies in the Yellow Pages, probably under 'Careers' or 'Employment Agencies'. Find out exactly what a firm is offering before committing your money.

● *Forty Plus Foundation.* The foundation is a non-profit organisation sponsored by leading companies in the UK and by the MSC. Six months membership costs £300, and is often paid as part of a redundancy deal. Members are selected on the basis of good careers and potential. The foundation provides a short course on self-marketing and jobhunting techniques, as well as secretarial services, telephones, copying facilities and reference books. Members work together to overcome the double disadvantage of being unemployed and over forty. Information from Templar House, 81/87 High Holborn, London WC1V 6NP.

● *Executive Stand-by.* Executive Stand-by is a specialist organisation which deals primarily (though not exclusively) with: Experienced executives and directors. People aged 50 and over. Short contract employment.

The role of Executive Stand-by is to bring together employers who are in need of short-term, temporary or long-term contracts and people with the experience and skills to meet the needs. They also include voluntary organisations on their books, as well as firms from industry and commerce. Information: Executive Stand-by (310 Chester Rd, Hartford, Northwich, Cheshire CW8 2AB, tel: 0606-883-849), or Executive Stand-by (Midlands) Ltd (60 Beaks Hill, Kings Norton, Birmingham B38 8BY), or Executive Stand-by (NE) (37/39 Victoria Road, Darlington, Co. Durham) or Executive Reserve Manpower Services Ltd (Templar House, 81/87 High Holborn, London WC1V 6NP).

- *The Institute of Marketing* (Moor Hall, Cookham, Maidenhead, Berks) offers a guidance and counselling service to its members; a detailed redundancy kit which includes various lists of head hunters, recruitment agencies and consultancies; and personal advice on redundancy and on marketing yourself.
- *The Birmingham Job Change Project* is a new and bold project, organised under the auspices of Birmingham Polytechnic. As well as helping to direct people through to new jobs, or to new training or retraining, the project is developing the possibilities that people might be given help to move into self-employment (Polytechnic, Stratford Road, Birmingham).
- *New directions*. There are a number of initiatives being developed around the country, in response to the crisis, and to the personal straits that people find themselves in. The Association of Teachers of Management is co-ordinating a working group, in order to spread ideas, and to inform people about new approaches to the problem of redundancy. They hope to be building up an information bank of projects, self-help groups, etc. For information, write to John Appleyard, ATM, Polytechnic of Central London, 35 Marylebone Road, London NW1 5LS.
- *Voluntary Services Overseas* places skilled people over 21 in voluntary positions in developing countries outside Europe. Fares and a living allowance are paid. They are keen to take volunteers with skills in education, health services, agriculture, engineering, crafts and trades, and social or business development.

  Details from VSO, 19 Belgrave Square, London SW1X 8PW.
- You could also write to *International Voluntary Service* for their current vacancy list for skilled volunteer posts overseas. They seek people who:
  - Have a vision of a world-wide community based on justice, equality and co-operation.
  - Want genuinely to assist those disadvantaged by international competitive and exploitative structures.
  - Have useful skills, training and experience to use and pass on.

  Write to Section OS54 IVS (HB), 53 Regent Road, Leicester LE1 6YL.
- *Colleges and universities*. Most colleges and universities have their own careers appointment services. If you are a graduate, or have a college degree of any kind, you can use the careers service of your nearest college or university.
- There are also specialist professional associations, particular to different lines of work whose addresses can sometimes be found in the Yellow Pages under Trade Associations and Professional Bodies.

Books

How To Get a Job, by Marjorie Harris (Institute of Personnel Management).
Getting the Job You Want, by Howard Dowding and Sheila Boyce (Ward Lock).
Finding Another Top Job, by Bill Lubbock (Institute of Personnel Management).
Changing Your Job, by Godfrey Golzen and Philip Plumbley (Kogan Page Ltd.).

Chapter 7: Education and training

INTRODUCTION — HELP WITH THE BASICS — A TRAINING IN A SKILL — TO MOVE INTO HIGHER EDUCATION — RESIDENTIAL ADULT COLLEGES — PART-TIME STUDYING — DOING A COURSE FROM HOME — FINDING OUT MORE — A FEW WORDS OF FRUSTRATION — A GENERAL WORD OF ADVICE

INTRODUCTION

'After being stuck without work for four months, I was just beginning to rot away in my own boredom. I thought, "Right, this is no good. I'm going to find out everything I can about courses, and studying, I've had enough of going silly at home." I didn't know much about it all to begin with. I called in at the Technical College, and asked to speak to someone about courses.

'Well, by the end of that week I had joined an evening class in home wine-making, and had started thinking about doing a couple of 'O' level courses, to get my rusty brain back into gear. What really shook me, and changed my life, you could say, was when I learnt that as an adult, or "mature student", I could apply for a place at a university or polytechnic, for a degree. A degree! I ask you. I'd never dreamt that I might be able to do that kind of thing. I didn't have any 'A' levels — but I was told it didn't matter.

'I had an interview at Hull University last month — quite a long one — and I had to write a paper for them, and they accepted me! I'm starting a three-year degree course in geography in October, and I'll be getting a full grant to support myself and my family. It's going to make a big difference to my life.'

(Driver, aged 29.)

Did you know that:

You can be paid to learn or train (through TOPS courses and various grants)? ☐

You can go to university without 'O' or 'A' levels (there are many special entry schemes for mature students)? ☐

You can do a degree without going to a university? ☐

You can spend up to 21 hours a week studying at college, without your right to benefit being affected? ☐

Most evening classes are either half-price or free for unemployed people? ☐

You can get a part-time course started specially for you, if there are several of you? ☐

There are eight residential colleges just for adults, where you can study on a grant for one or two years? ☐

You can get help to improve your reading, writing and maths, usually for free? ☐

You can study at home by a correspondence course, radio or TV programme? ☐

Although cuts are being made, there is still a huge variety of different kinds of course available? ☐

There are people whose job it is to help you find out what is available? ☐

Why bother?

- It will keep you occupied, and give you something to do, instead of doing nothing. ......

- When you apply for a job, or go for an interview, it will *show* that you've been keeping occupied, which will be to your advantage. ......

- It will bring you new ideas and stimulation, and perhaps lead you into new, unexpected directions. ......

- You will meet new people and find yourself in new situations. It is possible that one of these people will indirectly lead you to a job. ......

- You could gain a new skill or a qualification. ......

- It *might* help you to get a job at the end (but it *can't* guarantee you a job). ......

Total score ...... (Max = 30)

Give yourself a score for each of the reasons above, where:

5 = a very good reason.		4 = a good reason.	
3 = quite a good reason.		2 = maybe a good reason.	
1 = not much of a reason.		0 = not a good reason.	

What do you think about your total score? Do you think it shows that there are probably enough good reasons for you to read on, to find out more? Yes ☐ No ☐

What kind of thing do you think you might want?

Tick if interested

☐ 1. Help with the basics — improving reading, writing and numbers skills.	Adult literacy and numeracy schemes, and TOPS Preparatory courses.
☐ 2. A training in a skill.	TOPS courses and vocational courses at colleges.
☐ 3. To move into higher education.	Full-time study on a grant at a university, polytechnic or college of higher education.
☐ 4. To spend one or two years at a residential adult education college.	There are eight such colleges in the country.
☐ 5. To study or learn something on a part-time basis, at day or evening class, using the '21 hours' ruling.	There are classes available in everything from car-maintenance to bee-keeping, or 'O' and 'A' levels.
☐ 6. To do a course from home.	There are correspondence courses and FlexiStudy courses you can do.

HELP WITH THE BASICS

'I was rather embarrassed about my problems over reading and writing, and didn't like to talk to anyone about them. Then I saw the television programme *Write Away* and got in touch with the phone number they gave out. I'm getting regular help now at an evening class. The tutor's been doing things like application letters and forms too. It's good to know I'm not alone.'

(Unemployed storeman, aged 39.)

You can find out about 'Basics' groups, classes or individual help in your area either by contacting the Adult Education Dept. (p. vii) or by phoning one of these numbers. They will pass your name on to the local organiser:

England: 01-992 5522;   Wales: 022-869444;
Scotland: 041-332 4028.   N. Ireland: 0232-22488.

TOPS do short *Preparatory courses* in basic skills, designed to help people whose standard of reading, writing and arithmetic makes it difficult for them to get a job, and to prepare people for the pre-entry tests for other TOPS courses. You would get a full weekly TOPS allowance. Details: at the Jobcentre, or phone the local TSD office (p. viii).

Keep an eye open for TV programmes like *Write Away* or *Your Move* that may help.

A TRAINING IN A SKILL

TOPS courses

These are government-sponsored training courses, run at TOPS Skillcentres and colleges. They are all full-time, and usually last for six months. If you get a place on a course, you will receive a weekly allowance. There is a very large range of courses available, but you will have to find out what is available locally.

Anyone can apply to go on a TOPS course, if these conditions are all met:

- You are aged 19 or over, and have been away from full-time education for at least two years.   Yes ☐   No ☐
- You intend to take up employment using the skill for which training will be given.   Yes ☐   No ☐
- You have not been on a government training course in the last three years. (This may not apply if you want an advanced course in a subject you have previously studied.)   Yes ☐   No ☐

How to find out full details

Ask at the Jobcentre if you can talk to one of the Employment Advisers about possible courses that might suit you. Leaflet TSD N100 (*Train for a Better Job with TOPS*) will tell you what courses are available, and leaflet TSD L91 will give you up-to-date details about the tax-free weekly allowance, the extra allowances for dependants and travel and lodging costs, if you need to live away from home while on a course. It might help you to decide what to do if you make an appointment by phone to talk to someone at the Skillcentre.

TOPS run courses in engineering, construction skills, electrical skills, office skills, automative skills, management skills, hotels and catering, HGV licence, starting up a small business, higher level courses, and miscellaneous skills (computer programming, stock-keeping, hairdressing, etc).

How to get onto a TOPS course

You can apply at any time, bearing in mind that there is usually a waiting list (perhaps several months long). You will be asked to do a short maths test. If you fail this, ask straight away about the TOPS Preparatory Courses (see above), and apply again when you have been through one of these. For some courses there may be a second test, too. If you pass the tests, you will meet an interview panel, who will want to find out how serious you are about training, and about working at the trade in question when you have finished the course.

The courses are quite tough, with strict attendance hours (often 8.00 to 4.45).

About 50% of the people who finish TOPS courses do end up in jobs that use their new skill.

If you are accepted, and have time to wait — use Chapters 2 and 8 to help you plan your next few months, so that you make the best use of your time.

TOPS 'Infill' Course Sponsorship

TOPS can also pay for people to go on other training courses held at colleges or other places, so long as they don't last more than a year, and are *vocational* — they prepare you for a job. (Not 'O' or 'A' levels.) If you know about a course that you want to do and you think you might be able to get TOPS sponsorship, apply to them immediately. All applications are treated on a 'first come, first served' basis. You might be able to get sponsorship for courses such as horticulture, agriculture, hotel and catering, secretarial, diploma and certificate courses, crafts courses such as pottery, thatching, picture-restoring, enamelling.

What should you do to follow this up? Find out about courses in the area that interests you. See below — 'How to find out more'.

TOPS Wider Opportunities Courses

These are open to anyone over 16, and designed for people who are not sure what kind of work they want to do. They last about 12 weeks, and experience of various kinds of skill is given. Normal TOPS allowances. Details through the Jobcentre.

(See also Chapter 6, 'Wider Opportunities for Women' and Chapter 9, Small Business courses.)

Vocational courses at colleges

There are many colleges, called variously 'Technical College', 'College of Further Education', etc. They all run courses that are vocational, which teach practical skills, but the trouble is finding a way to do a course without having to pay a lot.

As mentioned above, TOPS 'Infill' sponsorship is one possibility.

If you are aged under 19, then the local authority will pay, and perhaps also give you a very small grant, to help with living costs.

Otherwise, the '21 hour' rule (see the section on 'Part-time studying') may allow you to study for up to 21 hours a week, without your benefit being affected. As for the cost of the course itself, it may be free or half-price for unemployed people. Since many courses are arranged for students who are in jobs, on a part-time 'sandwich' basis, it might be possible for you to attend. Visit your local college, and arrange to have a talk with someone who will tell you what courses are available. See *Second Chances for Adults*, details below.

TO MOVE INTO HIGHER EDUCATION

Here we are talking about studying for a certificate or a degree at a university, polytechnic or a college of higher education. This may not be as impossible as it sounds, because once you are an adult, or a 'mature student', you can apply for entry without having the usual 'A' levels that school-leavers have to have. On the whole, since the field is so large, it is rather beyond the scope of this Handbook. If you read the book *Second Chances for Adults*, you will be able to get a very full picture of the kind of opportunities that are available (details below).

As for grants, if you are accepted onto a degree course (and some other higher educational qualifications) you will automatically get a grant, so long as you've not had one before. For other qualifications, the grants are only 'discretionary', and your local authority can decide whether it wants to give you one or not. Policy varies a lot from one county to another. You can buy the *Grants Survey* from the NUS (3 Endsleigh Street, London WC1, £2.50), and the booklet *Educational Charities* (£1.00).

RESIDENTIAL ADULT EDUCATION COLLEGES

There are eight of these colleges in the country. They are all for adults who want to take a one- or two-year education (not training) course, mostly in the humanities, social sciences, industrial relations and trade union studies. Acceptance is by interview, not by qualifications, and if you are accepted there is a standard procedure for getting a full grant. Their standards of academic education are of a very high quality.

To find out more, write for information to:

The Registrar, Coleg Harlech, Harlech, Gwynedd LL46 2PU.

Director of Studies, Cooperative College, Stanford Hall, Loughborough, Leics.

The Bursar, Fircroft College, Selly Oak, Birmingham B29 6LH.

The Secretary (AE), Hillcroft College, South Bank, Surbiton, Surrey KT6 6DF.

The College Secretary, Newbattle Abbey Adult College, Dalkeith, Midlothian.

The Registrar, Northern College, Wentworth Castle, Stainborough, Barnsley, South Yorks S75 3ET.

The College Secretary, Plater College, Pullens Lane, Oxford OX3 0DT.

The General Secretary, Ruskin College, Oxford OX1 2HE.

PART-TIME STUDYING

The local college, the adult education department, the workers' education authority (WEA) and university extra-mural departments all arrange classes which take place both in the evenings and during the day. If you go to the local library and ask for details about each of these four organisations, you should have a good idea of what kind of class is available. You can then take your choice, from '19th century novelists' to 'O' level French at the academic end of the scale, or from welding to batik dyeing, at the practical end. As is stressed in Chapters 2 and 8, this kind of studying can be valuable for keeping you active and demonstrating to an employer that you have made the most of your time. If you can go with a friend, it'll make it more fun.

The classes may be free, or at least half-price, for unemployed people. It may cost you something to get there, so a push bike is invaluable. Remember that you are allowed to earn up to 75p a day on Unemployment Benefit, and £4 a week on Supplementary Benefit.

Many colleges now run special 'return to work' and 'return to study' classes, which may well be worth asking about. These prepare people for full-time work or study, and help to get the rust off.

The '21 hours a week' rule

This '21 hours a week' rule was originally outlined on Administrative Memorandum 4/77 from the Department of Education and Science. What it says is that instead of having to be constantly 'available for work', unemployed people can attend college courses in the day-time without having their benefit cut off, as long as they can leave the course as soon as a job comes up. But your office may not know this, so you may have to stand up for your rights and appeal.

DOING A COURSE FROM HOME

If you find it difficult to get to a college (Is it too far? Are the courses at the wrong time? Doesn't the college offer the course you want?) you may be able to study at home.

Write to the National Extension College (18 Brooklands Avenue, Cambridge CB2 2HN) for a free brochure about the many correspondence courses that they run. NEC also runs a 'FlexiStudy' scheme, whereby a number of local colleges will allow you to do a correspondence course with help from staff at the college.

If you are interested in doing a degree from home, write to the Open University (Walton Hall, Milton Keynes MK7 6AX) to find out details about their degree courses. You don't need any 'O' or 'A' levels to do an OU degree.

For details of other correspondence courses, write to the Council for the Accreditation of Correspondence Courses, 27 Marylebone Road, London NW1 5JS.

FINDING OUT MORE

The great variety of further education and training possibilities can make it all seem a bit complicated. So here is the basic advice if you want to find out more:

Get in touch with your *local college*, and arrange to talk to someone there about further education in general. He or she will probably be able to tell you many things you didn't know. Phone first, and make an appointment; you'll get much better advice that way.

In many areas there are specialist *educational advice services*. The local Citizens Advice Bureau ought to know if there is one locally, so phone and ask. Or write for a complete up-to-date list to Michael Redmond, Senior Counsellor, The Open University, Fairfax House, Merrion St., Leeds LS2 8JU (0532–44431).

Visit the *Careers Service*, where there are trained Careers Officers whose job it is to know all about education, qualifications, entry requirements for jobs, etc. Phone first, to make sure you get a proper appointment.

Second Chances for Adults by Andrew Pates and Martin Good (Macmillan Papermac £3.95) is an invaluable and comprehensive guide to the opportunities available. It is updated every year, the new edition coming out in September. If your library hasn't a copy, ask them to get one. This will answer most of your questions, and point you in the direction you want for more detailed information.

Fresh Start — A Guide to Training Opportunities comes *free* from the Equal Opportunities Commission, Overseas House, Quay Street, Manchester M3 3HN. A brief guide to opportunities, written for women, but containing useful addresses for everyone.

A FEW WORDS OF FRUSTRATION

It would not be right to let this chapter pass without some comment on the cutbacks taking place in education. It's crazy — a time of high unemployment and of rapid social change has to be a time for big *expansion* of all education and training facilities, in preparation for the future. Even evening classes are being cut back in many areas. The lack of any well-supported educational response to the unemployment crisis, in the form of new courses, new learning arrangements and extra finance for grants and courses is very depressing. It's one extra thing to fight for — see Chapter 12.

A GENERAL WORD OF ADVICE

If someone says to you, 'It isn't possible', 'There aren't any places left' or, 'She's too busy to see you', *persist*. Don't just walk away. The variety of courses available is large, and it is quite common to find that people at one end of a college don't know what's happening at the other end of the same college. Also, it's just life that some people seem to enjoy telling people that it *can't* be done, whereas others enjoy trying to bend every rule so that it *can* be done. Go on till you find one of the second kind. Remember, too, that the higher up you go, the more power people have to make exceptions to their own rules. So — *persist*.

Chapter 8: All your own time

INTRODUCTION – REBUILDING SUPPORTS – GETTING DOWN TO IT – CATALOGUE OF ACTIVITIES – MAKING THE MOST OF YOUR COMMUNITY – SPORTS – HOBBIES AND PERSONAL INTERESTS – DIY REPAIRS – EARNING MONEY – GROWING YOUR OWN FOOD – SETTING UP A FOOD CO-OP OR BULK BUY CLUB – A CRAFT CO-OPERATIVE – WORKING WITH CHILDREN – TEACHING YOUR OWN CHILDREN – GETTING INVOLVED IN A TENANTS ASSOCIATION – VOLUNTARY WORK IN YOUR NEIGHBOURHOOD – SHARING AND DEVELOPING SKILLS – ACTION GROUPS – PERSONAL CARE AND COUNSELLING WORK – LOCAL ACTIVITY AGAINST UNEMPLOYMENT – INFORMATION AND ADVICE – ACTION

INTRODUCTION

If you want to do things that will bring you satisfaction and give you experience, the catalogue in this chapter may give you some ideas. You may find it helpful to think back to Chapter 2, where we looked at the supports we need when out of work, and then to Chapter 3, where you made up a list of what your own natural skills and abilities are.

REBUILDING SUPPORTS

The supports we mentioned in Chapter 2 were:
- Purpose and direction.
- Regular daily activity.
- Identity and self-respect.
- Friends and colleagues.
- Money.

It will definitely help you in the section below if you read through the 'Catalogue of Activities' first, and then return to this.

1. Purpose and direction

Make a list of activities that might bring a sense of purpose. (Jobhunting itself? Learning a new skill? Involvement in a neighbourhood group?) Choose two that appeal to you most, and write them below.

1. . . . .   2. . . . .

2. Regular daily activity

Make a list of activities that you could do on a regular daily basis. Choose two, and write them below.

1............................   2...........................

....

....

3. Identity and self-respect

We can gain a firm sense of identity and self-respect if we have a chance to prac-tise our natural abilities. Chapter 3 may help. If you haven't done the exercise there on finding your natural skills and abilities, do it now. Then make a list of these skills and abilities.

Now think of several activities that could give you a chance to express these skills and write two here.

1............................   2...........................

....

4. Friends and colleagues

Make a list of activities that could bring you into regular contact with people, and then write the two that appeal most below. If you can't think of enough, look through the 'Catalogue of Activities' again and see which ones look interest-ing.

1............................   2...........................

....

5. Money

There may not be too much you can do here — but do check Chapters 9 and 10. We'll add another important support instead — physical health.

6. Physical health

Make a list of any activities that may help to keep you fit, and then choose two that you would like to do:

1............................   2...........................

....

GETTING DOWN TO IT

Step A

Take all the activities you have written down or are interested in and ask these questions about each one, using a separate sheet of paper for each activity:

1. Am I serious about wanting to do this? Yes ☐ or No ☐ ?

2. How am I going to go about it? Who could help or advise me?

3. Are there any problems or difficulties associated with it?

4. How could I overcome them? Who could help or advise me?

5. Am I really going to press ahead, and do this? Yes ☐ or No ☐ ?

Repeat this process for each activity, and then move on.

Step B

Take a whole page in your Journal, or a sheet of paper, and draw up a diary for a week filling in all your activities.

Step C

Find a friend, preferably one who is in the same situation as yourself, and show him or her what you are doing. You may be able to do some things together; even if you don't, it will be helpful to keep in touch with each other's progress. If you are a member of a regular unemployment support group, then do the same thing with them.

It is *not easy* to live a self-disciplined life without any factory or office hours telling you when to work and when to stop. You will be unlikely to achieve *all* your good intentions. But don't give up altogether. Just remember that what you are trying to do is hard, and that other people have just the same problem.

Step D

Move into action!

CATALOGUE OF ACTIVITIES

Making choices

After each entry there is a place for you to record how much the activity interests you, using a scale from 0 to 10, where:

10	=	definitely want to do it.	4	=	quite interested.
8	=	very interested.	2	=	not very interested.
6	=	definitely interested.	0	=	not interested at all.

1. MAKING THE MOST OF YOUR COMMUNITY

A good starting point is to find out just what is going on in your area. Use this questionnaire to check what facilities there are.

Local Community Resources

Does your local neighbourhood or area have:

A festival ☐
Enough playgroups ☐
A youth club ☐
A drama group ☐
A choir or chorus ☐
A dance group ☐
A good choice of evening classes ☐
Facilities for football ☐
Facilities for tennis ☐
Facilities for badminton ☐
A swimming pool ☐
Bowls ☐
Areas for play safe from traffic ☐
An information/advice centre ☐
A 'Small Industries Group' (see p.141) ☐
A 'Young Enterprise' group (see p.143) ☐
A cinema or film club ☐
Scouts, guides, brownies, cubs ☐
A local rock group ☐
A folk club ☐
A print workshop ☐
A community magazine ☐
Clubs for the disabled and/or house-bound ☐
Clubs for the mentally subnormal ☐
Decent local transport arrangements ☐
A holiday playscheme ☐
Enough allotments ☐
A neighbourhood law centre ☐

A women's group ☐
A local environment concern group ☐
A library ☐
A decent hall ☐
Meeting rooms ☐
A credit union ☐
A street warden scheme ☐
A housing action campaign ☐
A street market ☐
A parent-teachers association ☐
A 'Gingerbread' group for single parents ☐
A schools action group ☐
An unemployment support group ☐
Good local dances, discos ☐
Regular public meetings to discuss local problems ☐

. . . . (add extra)

. . . . (add extra)

TOTAL ☐

You may be interested to see how your area scores:

- 30-35: Seems like your community is alive and well — so you shouldn't have much difficulty finding things to do.
- 20-30: Your community has woken up, and is starting to do things.
- 10-20: It's just about got the sleep out of its eyes.
- 0-10: Still pretty sleepy.

But even if your area isn't one of the liveliest, you may be able to help it wake up. At Craigmillar, a large housing estate on the edge of Edinburgh, unemployment runs at four times the national average. Twenty years ago, it had nothing; then a few people set up the Craigmillar Festival Society, in the hope that it might start something off. It did. Youth clubs, playgroups, clubs for the housebound and disabled, lunch clubs, a community ambulance, an annual home-produced rock musical, an information and advice centre with a special job-finding service all started. And this has at last led to more jobs — a community owned and run company that hopes to be in business in two years.

So it can be done! Your interest in this kind of activity:

Score . . . .

See 'Information and advice', at the end of the chapter.

2. SPORTS

There should be a list of all the sports clubs which are active in your area in the library. If not, ask the Youth Service or the Adult Education Department (See Chapter 7 and p. vii.)

Which particular sport(s) interests you?

. . . . Score: . . . .

. . . . Score: . . . .

If there is a sport which you already practise, could you give coaching to younger people? If this interests you, call in at the local youth club, or phone the youth service to talk about it.

Which sport ? . . . . Score: . . . .

3. FOLLOWING A HOBBY OR PERSONAL INTEREST

The library should have a list of local clubs and groups. If you want to learn a new skill, or develop an existing one, you could find out about evening classes (see Chapter 7). There are also books on most subjects in local libraries. You could ask around to see if someone might be willing to teach you, perhaps on an exchange basis of some kind. You could also ask around (or advertise on a local notice-board) to find other people who want to learn that particular skill.

Which hobby/interest ?. . . . . Score: . . . .

. . . . Score: . . . .

As above, you might consider sharing your interest with younger people in the area.

Which interest ? . . . . Score: . . . .

4. DO-IT-YOURSELF HOUSE REPAIRS

This might be a matter of financial necessity, as well as a chance to learn various skills. Use the library, evening classes and advice of friends. See *The Self-Help Housing Repairs Manual*, by Andrew Ingham (Penguin 60p). Join together with your friends, to help each other.

Score: . . . .

5. EARNING MONEY

The rules about earning money while unemployed are in Chapter 10. People *do* build up small businesses from very small beginnings, and anyway, there are obvious reasons why it could be useful to have a little extra cash coming in, even if it is only £4 a week.

Score: . . . .

6. GROWING YOUR OWN FOOD

If you have a garden but you don't grow food in it yet, you will find that other gardeners just *love* giving advice. Local gardening shops are also good sources of advice, and libraries have books on gardening. You could save (and perhaps even earn) money and have your own fresh vegetables to eat.

Growing food and selling on Sundays.

What if you have no land?

- Apply for an allotment to the parks department of the council, or the parish/town council. There will probably be a waiting list, but there may be a vacancy. There are sometimes private allotments around, too, which other gardeners may know about.
- Look around for unused land, owned by the railway, the Gas Board, the Church Commissioners, developers, etc. and then get together with a few friends to ask if you could rent it cheaply on a short-term year-by-year lease. If they say no, write to the press, raising the whole question of why people are allowed to leave good land lying around idle when it could be in productive use.
- Friends of the Earth operate a garden-sharing scheme, by which people who have gardens that are too big for them allow others to use part of them. The booklet *Crops and Shares* explains how it works. (Price 30p + 15p postage from FOE, 9 Foland St., London W1V 3DG.)

Score: . . . .

Getting experience of growing food/horticulture

A group called 'Working Weekends on Organic Farms' enables people to spend a weekend helping on someone's farm or smallholding. There is no pay, but you get free food for the weekend, a bed, and pleasant company. To get the details, you have to join WWOOF, which costs £2, for which you receive their regular newsletter giving the addresses and details of farms wanting help. (WWOOF, 19 Bradford Road, Lewes, Sussex.)

Score: . . . .

7. SETTING UP A FOOD CO-OP OR BULK BUY CLUB

It is possible to get together with other people and buy your food and supplies direct from wholesalers, farmers, markets and cash-and-carries, which makes things cheaper. There are many such co-ops or clubs around the country working on a voluntary basis. For up-to-date advice, send £1 for a copy of *The Bulk Buy Book* to the National Consumer Council, 18 Queen Anne's Gate, London SW1H 9AA. (The book *People Power*, see below, also includes advice.)

Score: . . . .

8. A CRAFT CO-OPERATIVE

If you do any kind of craft, you have probably met the problem of how to sell the things you make. The Aguirre Craft Co-operative in Brixton had the same problem; they set up a shop together (which they ran collectively) to sell the craft goods that they and their friends make. For details, see the information sheet list at the end of this chapter, and the final section of Chapter 9.

Score: . . . .

9. WORKING WITH CHILDREN

- Helping in *playgroups* for the under 5s, where offers of help from both men and women are always welcome. You don't have to be skilled or trained — you can pick skills up as you go along from the other helpers. The Social Services Department keeps a list of local playgroups.
- Working with *holiday playschemes* and *adventure playgrounds* for 5–11 year olds and 11 – 16 year olds. This can involve going off in minibuses for trips, refereeing games of football, playing with inflatables, organising street and party games, making candles, teaching unarmed combat, etc. Every adult has something they can offer. The Social Services Department will know about local playschemes, or write to the National Playing Fields Association, 25 Ovington Square, London SW3.

Score: . . . .

- Helping in *youth clubs*. Many youth clubs seem to have very few activities beyond table tennis, billiards and discos. These are always popular, but so are caving, rock-climbing, swimming, youth-hostelling, camping, putting on a play, turning wasteland into somewhere exciting and pleasant, hiring a barge for a canal outing, laying on an evening's entertainment for the elderly, printing your own T-shirts, etc. All these things need people to organise them. The Youth Service will know where local youth clubs are.

    See also *KIDS! — A Handbook For Those Working With the Under 14s* (price 50p from National Association of Youth Clubs, PO Box 1, Blackburn House, Bond Gate, Nuneaton, Warwicks CV11 4DB; full of games, ideas, etc. either for the family or for playschemes, youth clubs, etc.).

Score: . . . .

10.   TEACHING YOUR OWN CHILDREN

Some parents prefer to educate their children themselves, at home. It is not an easy thing to do, but you can get support, if you want to try, from a group called 'Education Otherwise'. They have produced a well researched guide, based on experience: *Education Otherwise. . . . . First Steps.* 40p from Education Otherwise, 69 Leathwaite Road, London, SW11.

Score: . . . .

11.   GETTING INVOLVED IN A TENANTS ASSOCIATION

In many cities and towns local residents form tenants associations in order to make their complaints felt, get repairs done, try to prevent vandalism, and generally make the area a better place to live in.

Becoming involved in such a group can teach skills such as how to run a meeting, how to confront the council, how to organise a petition, how to speak confidently in public, etc.

If there is no tenants association in your area, but you think that there should be, you can get advice on how to form one from Shelter (157 Waterloo Road, London SE1 8UU) and from the *Grapevine* sheet listed at the end of the chapter.

Score: . . . .

12.   VOLUNTARY WORK IN YOUR NEIGHBOURHOOD

There is a constant need for people to help out in hospitals, in homes for the elderly and the handicapped, in children's homes and in day-centres. Just going in for one afternoon a week to a children's home to read stories may not seem much, but if it is done regularly it means a lot to the children. The same applies to helping out once a week at a senior citizens luncheon club, etc.

In most areas there is a body called the 'Council for Voluntary Service', or the 'Volunteer Bureau', which keeps information about who needs help, and who is offering help. They will be able to tell you where help would be appreciated locally, and put you in touch. (Address, see p. viii.)

Score: . . . .

Voluntary work is not all of this kind — there are usually groups saving old railway lines, cleaning up old canals, restoring wasteland, keeping checks on rare species of plant, etc. You will be able to find details of these groups in the library, and through the Council for Voluntary Service.

Score: . . . .

13.   SHARING AND DEVELOPING SKILLS

Everyone has skills of different kinds, which do not need to go idle just because you are unemployed. If one unemployed person is a plumber, another is a painter and another a good typist there are plenty of ways in which they can help themselves and others without any money having to be involved at all. To develop this kind of sharing, 'Skills Exchange' projects are beginning to spring up.

As one example, 'Rally Merseyside' has been set up by the Merseyside Council for Voluntary Service (Inner Temple, Temple Lane, Liverpool L2 5RS) 'to enable people who have time to give to make use of their time, skills, occupations and other goods and services by exchanging them'. People phone the network, making either an offer or a request. The network team matches offers with requests, and an exchange is made. No money is involved. It is run from a basement in a backstreet in central Liverpool. (Rally Merseyside is also seeking to develop an experimental centre to help and support people wishing to take on community projects, to provide resources, equipment, training and education and to study the whole field of alternative work in the urban setting.) For information on how they do it, and advice, just write to them.

Score: . . . .

14.   ACTION GROUPS

In most areas there are pressure groups, action groups and campaigns fighting for different kinds of change. Friends of the Earth groups work away at recycling paper, saving whales, stopping pollution, fighting the nuclear energy programme, planting trees, etc. (9 Poland Street, London W1V 3DG and 69 Cumberland Street, Edinburgh). Christian Aid groups campaign for Third World development (PO Box 1, 2 Sloane Gardens, London SW1). National Council for Civil Liberties groups try to preserve and protect our individual liberties against encroachment by police, state and computers (186 Kings Cross Road, London W1). Campaign for Nuclear Disarmament groups campaign against nuclear weapons and the threat of world nuclear war (CND, 29 Great James Street, London WC1).

Which group? . . . .   Score: . . . .

15.   PERSONAL CARE AND COUNSELLING WORK

This is another kind of work which is rather different. It is a way of helping other people by being there and listening when they need help. Some people are 'easy to talk to' and this can have a positive effect. Many people do this among their friends anyway. It is also possible to put such skills to use in a more organised way, through one or other of the groups mentioned below:

The Samaritans, 17 Uxbridge Road, Slough, Bucks. The address of the local group will be in the phone book.

Marriage Guidance Council (relationship counselling), Herbert Gray College, Little Church Street, Rugby, Warwicks.

Natural Health Centres are voluntary organisations which inform people about natural ways of healing and local therapists. Some centres provide a counselling service for people who want someone to talk to, and offer relaxation sessions. You can get a list of these centres by writing to the Healing Research Trust, c/o Totnes Natural Health Centre, 69 Fore Street, Totnes, S. Devon.

Which group? . . . .   Score: . . . .

16.   LOCAL ACTIVITY AGAINST UNEMPLOYMENT

Chapter 12 has details about a number of ways in which various groups have begun to tackle unemployment. The problem itself could become the origin of changes which have a profound effect on the way we live and work, and the way we organise our local economies. In this kind of work, there is ample outlet for skills and abilities, and a real opportunity to give valuable service to the community. Here is a brief list of different ways in which it is possible to tackle unemployment:

- Set up support groups for the unemployed.
- Set up benefits advice centres.
- Campaign for changes in the rules and regulations about benefits.
- Establish a skills exchange for unemployed people, so that they can benefit from each other's skills and abilities without infringing the regulations.
- Set up/help with small industries groups.
- Hold public meetings to air local problems.
- Work on trade unions, to persuade them to include unemployed people in their decision-making processes.
- Campaign for an 'open information' policy for local firms, so that local people know the true facts of what is happening, and can take appropriate action *before* redundancies are declared, where this might help to save a firm.
- Campaign for extensions of the Youth Opportunities Programme, so that school-leavers all have a guaranteed place of work, training or education.
- Help set up community co-operatives and community-based enterprises so that people can begin to own and control the means of work within their own neighbourhoods.
- Persuade firms who are declaring redundancies to put money into the community for local projects, etc.
- Once a group starts to meet together to think about what can perhaps be done, there are really very few limits. It's really just a matter of deciding, 'It is time something was done.'

Score: . . . .

INFORMATION AND ADVICE

Once you know where to look to find answers to your questions, the business of trying to get something done becomes easier. When you begin to understand how the local council committees work, what the regulations are, how to get the press interested, how to arrange good publicity for a meeting, and who else is doing similar kinds of things, it becomes easier to achieve things. Having the relevant information, and knowing where to go to find things out makes you feel more confident. Here are some excellent sources of information and advice:

- The BBC programme *Grapevine*, which covers community self-help activities, puts out a number of superb information sheets on the topics covered in their programmes. Send a stamped addressed envelope to *Grapevine* (BBC TV, London W12 8QT) for any of the following information sheets: *Holiday playschemes, Fund-raising, Advice and information centres, Food co-operatives, Community festivals, How to use your local radio station, How to form a group, Child care, Photography, How to paint a community mural, How to transform a school playground, Share-a-garden schemes, City farms, Community publishing, How to make a record, Community print shop, Market stalls, One-parent families, Steel and skin — Afro Caribbean Workshop, Architectural and building advice, Rural self-help schemes.*
- *People Power — Community and Work Groups in Action*, by Tony Gibson (Penguin, £1.25). Part One of the book describes a number of projects where people have achieved local successes, and gives detailed case-studies.
  Part Two is called the 'Neighbourhood Fact Bank' and consists of several hundred units of basic, concise information, covering such things as setting up a playscheme, insurance, repair and rebuilding costs, traffic hazards, housing co-operatives, grants, lotteries, tenants' rights, council regulations, etc.
- *Getting Self-Propelled — A basic guide for self-help groups* (70p) and *Making the Most of Neighbourhood Skills and Experience* (50p) are two very useful publications from Education for Neighbourhood Change (c/o The School of Education, Nottingham University, Nottingham NG7 2RD), produced by the team that *People Power* comes from. Recommended for anyone involved in local community actions.
- *The Village Action Kit*, by Guy Dauncey and Brian McCloughlin, is a rural equivalent to the above, containing detailed information about rural self-help initiatives in the fields of housing, transport, local shops, unemployment and education. £1.95 from the National Extension College, 18 Brooklands Avenue, Cambridge CB2 2HN.
- *Meanwhile Gardens* (75p from Gulbenkian, 98 Portland Place, London W1N 4ET) is a first-hand report about how a group of people turned a scrap of wasteland near Paddington, London, into an exciting place including a cycle speedway, a garden and an adventure play area.
- *Docklandscape* by Hilary Peters (Watkins Publishing, Dulverton, Somerset, £3.00) tells how a small group of people turned parts of the Surrey Docks in the East End of London into a place with gardens, goats, poultry

and allotments.
- *The Fundraising Handbook*, by Hilary Blume (£2.15) is the best book on fund-raising, covering local events, getting grants, charities and foundations, raising money from industry, etc. and *The Directory of Social Change: Vol. 2: Community*, by Michael Norton (£3.95) covers a mass of community self-help information and ideas. Both from Directory of Social Change, 9 Mansfield Place, London NW3.

You can order any of these books through your library.

ACTION

Go back through the catalogue, and make a list of all the activities for which you scored 6 or more:

Score

_____

_____

_____

_____

_____

_____

_____

_____

_____

_____

Now go back to the beginning of the chapter, and work through the 'Rebuilding Supports' exercise, using the list of activities above to help you.

Chapter 9: Self-employment

INTRODUCTION — THE IDEA — HAVE YOU GOT THE SKILLS? — HOW
MUCH MONEY WILL YOU NEED? — PREMISES — RULES AND REGU-
LATIONS — PREPARATIONS — WERE YOU SERIOUS? TAKING THINGS
FURTHER — WHAT ABOUT A CO-OPERATIVE?

INTRODUCTION

There is an old cartoon joke about the street full of small tobacconist shops
opened up by friends who were all made redundant at the same time.

It is an old dream, being self-employed, which is deeply engrained in the
whole British tradition. Independence, no boss breathing down your neck, no
one telling you what to do or what hours to work. To be your own master.

But is it possible? Is it a realistic dream? In particular, is it possible or realistic for people who are unemployed, and probably broke?

This chapter has been designed as a game. You can play it whether or not you
are serious about becoming self-employed. It would lend itself well to being
played in a group. (See Chapter 5.) The game asks you to think of three possible
ideas for your own self-employment, and then it takes you through six stages of
questioning and planning, leading finally to a seventh stage when you decide
whether you want to take one of your ideas further. The main aim of the chapter is to show that once you closely examine a possibility for self-employment, it
might not seem so crazy after all.

STAGE 1: THE IDEA

What could you do? There are 1,001 possibilities. Some people take their craft
or hobby, and expand it until it brings in enough income. Some people think
more practically about what might bring in some good money and then decide
(for instance) to set up a taxi-service or to train to become a freelance plasterer.

- *Choose two possible ways* in which you think you might be able to become self-employed. If you need to, use the Yellow Pages to give you ideas
  — almost everyone who advertises there is self-employed. Think of skills
  that you have and things that you like doing. You do not have to start up
  straight away — you can give yourself time to train and get experience first.

  To be realistic, *think small*. Everything starts small — there will be
  plenty of time to expand later. Many new businesses collapse because they
  expand too quickly, before they have firm foundations.
- *Take three sheets of paper* (or three pages of your Journal) and on the

first sheet write down your ideas. Then draw seven columns next to them, as in Figure 13.

Take a sheet of paper for each idea, write the idea at the top, and as you move through Stages 2—7, make notes for each idea. The top sheet serves as a checklist, to be filled in as you work through the chapter.

	Idea 1	Idea 2
1 The ideas		
2 Skills (a) (b)		
3 Money (a) (b)		
4 Premises		
5 Regulations		
6 Preparations		
7 Taking it further		

Fig. 13. Your ideas

STAGE 2: HAVE YOU GOT THE SKILLS?

a) The actual skills for the job

If you think that you have already got the skills that you will need for one of your ideas, place a tick in Column 2(a).

If you do not have the skills, use the suggestions listed below to give you some ideas. Place ticks in Column 2(a) if you can decide how you could obtain the skills you will need.

Some examples

● *Making and selling hand-made toys.* Practising on your own, at home; evening or day classes in woodwork and wood-turning; going on a TOPS course in woodwork; getting hold of a wide range of toys, and taking them to pieces; going all-out to get a short-term job in someone else's toy workshop; asking to work in someone else's toy workshop without pay, while still signing on; offering to repair toys belonging to local playgroups, hospital children's wards, etc.; making toys to give away to such groups while you develop your skills.

- *Painting and decorating.* Practising in the houses of friends; going on a TOPS course; offering your services to voluntary groups in the area, especially to any involved in housing, or in helping the elderly, where there is always need for painters and decorators.

- *Running a private catering service.* Asking friends if you can lay on parties for them, if they provide the food; cookery classes at the local college, or at evening classes; getting cookery books from the library, and just trying out different recipes; getting short-term jobs in hotels; doing a TOPS Catering course; offering your services to voluntary groups in town which might be organising fund-raising events where catering is called for, e.g. special medieval feasts.

- *Book-keeping and secretarial services.* Going on a TOPS course, or a commerce or business studies course at the local college; private study and practice; offering your services to help with accounts and book-keeping for local voluntary groups.

- *Running a night club.* Going for short-term jobs in existing night clubs; getting to know all the people in the field, as friends and as contacts; offering to organise concerts, dances and events for local voluntary groups; organising a once-a-week folk/jazz or cabaret night at a pub.

- *Road Haulage.* Doing a TOPS course to get your Heavy Goods Vehicle licence, or doing it privately (around £200); working for an agency such as 'Manpower' as a freelance driver, or for an existing firm, before buying your own lorry; building up contacts in the field.

These examples should give you some idea of the varieties of ways in which you can build up skills and experience. Chapter 7 outlines details about training and learning; Chapter 8 outlines details about local voluntary groups. If you are under 19 you can add to each one above, 'arrange to do six months Work Experience (see Chapter 6) in that line of work'.

- *Write down how you would plan to get the skills you need, for each idea, on your sheets of paper.*

b) Business and organisational skills

Here we come to the matter of keeping accounts, getting orders, advertising, regulations, taxes, etc. — everything that is involved in being self-employed or in running a small business. There are various ways in which you can get these skills. (Some of your ideas may not need too many of these skills; others may need all of them.)

- Many adult education departments run *evening classes* in 'Starting a Small Business', or some such title. (Phone No. — see p. vii). If there are not any such classes, ask the person in charge if they could be organised. If the answer is 'no', try to find five or six friends who share your interest in attending a small business class (you could perhaps advertise ?), and write a collective letter asking for classes to be organised. This ought to be effective.

  A group of people attending an evening class in 'Starting a Small Business' is also a good basis for a 'Self-employment self-help group'. The idea here is very simple — you agree to meet together regularly to share your problems and to help each other in any way you can, with suggestions, ideas, contacts, etc. (See also Chapter 5.)

- 'TOPS' (see Chapter 7) run two types of course for training for self-employment: *New Enterprise Programmes*, for people who want to set up in a business which will employ a reasonable number of people; *Small Business Management Courses*, for people who want to start on a smaller scale. These courses are very successful, and are organised along exciting lines, with participants developing their own projects while on the course, aided by the staff. They last seven weeks, and are probably the best way to break into self-employment. Normal TOPS allowances. For details, ask for the booklet TSD N200, from the nearest TSD or PER office.

- Most colleges of further education have 'Business Studies Departments' which run a variety of courses. Phone up (see p. vii) to ask for details of their business courses, and discuss with them if there is anything that might suit you. You can either study in the evenings, or on a 21 hours a week basis while you are unemployed, or full-time on a grant. (See Chapter 7.)

  Some universities and polytechnics also have business schools, which the local college business studies department will know about.

- The National Extension College has produced a *Small Business Kit* (price £5.50 from NEC, 18 Brooklands Avenue, Cambridge CB2 2HN). The kit is designed to help you:
  - Decide whether self-employment is right for you.
  - Assess your capacity to run a small business and begin to develop it.
  - Work out whether a real opportunity exists for the establishment of the business you have in mind.
  - Assess your business idea in relation to your strengths and weaknesses, and the business environment.

- Other books:
  Working for Yourself, Godfrey Golzen (Kogan Page, £2.50).

Creating your Own Work, Micheline Mason (Intermediate Technology Publications Ltd., 9 King Street, London WC2. £2.30).
Occupation Self-Employed, Rosemary Pettit (Wildwood House, £2.95).
Going Solo (BBC Publications, 1975).
Setting up a Workshop (75p from the Crafts Advisory Committee, 12 Waterloo Place, London SW1Y 4AU).

All of these books can be obtained through your library, if you give them the full information. You can also obtain booklets from the Small Firms Information Service (free): e.g. No. 1, *Starting in Business.*

- The 'Small Firms Information Service' operates from centres all over the country, to give advice and help to people running small firms. They produce a wide range of free booklets on different aspects of starting and running a business, and they offer a free personal counselling and advice service.

  Freefone the nearest centre (in the phone book under 'Small Firms Information Centre', or 'Department of Industry'). The centres really exist to help established firms, and less to advise people who have not begun at all.

- For young people (under 25 or so), there is a scheme called 'Young Enterprise' which operates in many parts of the country. Young people learn business skills by actually producing a product, marketing it, and arranging the finance, distribution, etc. The groups meet one night a week over the winter months, and offer a very satisfactory way of learning how a business works from the inside. Some groups make quite sophisticated products — e.g. car burglar alarms.

  Write to Young Enterprise (46-47 Old Bond Street, London W1X 3AF) to see if there is a group in your area.

- *If you decide that by one method or another you could obtain the business skills necessary for your ideas, place ticks in Column 2(b).*

STAGE 3: HOW MUCH MONEY WILL YOU NEED?

a) To start

You will obviously need money to start up with, and to keep going on, until you start to bring in an income. What you need to do here is a quick thumbnail costing — just enough to tell you whether you will need £5, £500 or £5,000. Use the list below to work out your costs on each item:

- Tools and equipment.
- Materials (initial outlay).
- Stationery, advertising
- Transport.
- Cost of rented premises, and any alterations.
- Enough money to live on until you begin to bring in an income.

Do not forget that you can borrow, improvise, buy second hand and at auctions, etc. To give you some idea of costs:

- 200 sheets of headed notepaper   — £12
- 500 small trade cards   — £12
- Rent on a small workshop   — £12–£30
- Pocket calculator   — £4.95 + 3 Cornflakes tops.

- *Write your total in Column 3(a).*

b) Could you raise the cash?

Save up, over several months? Sell something? Approach a relative? Borrow? The Department of Health and Social Security has been known to make small grants to people on Supplementary Benefit to enable them to buy tools and equipment, if they think that this will enable someone to become self-employed after a while and sign off the dole. It's always worth a try. If you think you will need a bank loan, you will need to present clear facts and figures.

If you have got redundancy money, might this be one way of putting it to good use? Don't rush in, however. If you do not make careful plans, you could be throwing it all away.

- *If you think you could find the money you need, place ticks in Column 3(b).*

STAGE 4: PREMISES

No need to hang around on this one. If you think that you could start out based from where you live, place ticks in Column 4, opposite each idea. If you think you will need to find premises, seek advice from the council development office.

- *Place ticks in Column 4 if you feel confident about finding somewhere.*

STAGE 5: RULES AND REGULATIONS

You will need to handle your own *tax return* and keep *accounts,* or arrange for a friend who is good at it to come in and help. There are special account books which make it all quite easy to do, as long as you fill them in every day. You will need to pay your own National Insurance contribution (about £2.50 a week, and 8% of your profits once you earn over £1,750). You may need to take out *insurance* against liability, to protect yourself against any damages that you might cause. Any broker will advise you on this. You may also run into regulations like planning permission, fire regulations, food hygiene regulations, etc.

- *If you think that you could handle these things, place ticks in Column 5.*

STAGE 6: PREPARATIONS

Here are some of the things you will need to be thinking about, before you start up properly:

- Developing your skills.
- Developing your business ski ls.
- Iron out any technical problems.
- Make financial plans, and raise cash.
- Sort out about regulations, etc.
- Build up local contacts.
- Plan advertising, design, etc.
- Learn how to do accounts.

There are several 'routes' that you could take, all of which involve some delay before you begin. Take your choice:

a) *Continue to sign on* for a pre-decided period, and use the time to do everything you can to prepare, build up stocks, make plans, etc.

b) *Continue to sign on,* and declare your earnings from your self-employment each week. These will be deducted from your benefit (less a certain amount, see Chapter 10). When your income begins to exceed your benefit, sign off the dole (and throw a party!).

c) *For those on Unemployment Benefit only:* use the fact that you are allowed to earn as much as you like on Sundays, and that you can choose not to receive for one or two days a week, if you have a part-time job for those days.

d) *Aim to get a full-time job* (or work experience, for under 19s), and plan to build up your business slowly, in the evenings and at weekends.

e) *Apply for a place on a TOPS course*, in order to get the skills you'll need, and plan on a slow build-up.

f) Get a *part-time job*, or arrange to *share a job*, and use the other half of your week for your self-employment.

Don't rush in, or try to start before you are ready. If you want your self-employment to last, build yourself a solid foundation.

- *For both your ideas, mark one out of (a)—(f) in Column 6, depending on which route you think would be best.*

STAGE 7: WERE YOU SERIOUS? TAKING THINGS FURTHER

- *Take each of your ideas in turn,* consider them carefully, and place a final tick in Column 7 if you think that the idea is worth looking at a bit more seriously. It is unlikely that you will have more than one or two ticks in this column, if any.

To explore those ideas further:

- Find a sympathetic self-employed person to talk to. Try to get some realistic advice.
- Check through Stages 2-6 again, more rigorously.
- Use the NEC's *Small Business Kit* to develop your plans further. (See Stage 2(b).)

If you have got this far, it is time to give the warning that being self-employed is *not* easy. You get no paid holidays, and if you fall sick you will only get sickness benefit. There is only you yourself to make you get up each morning to work, even on the worst of days. You might also be on your own — unless you manage to find or form a self-help group with other self-employed people.

On the other hand, it will be *your* life, and *your* work. Good luck to you!

WHAT ABOUT A CO-OPERATIVE?

There are about 100 new small co-operatives being registered every year. One of these, which ran for several years in the late 1970s, was called 'Little Women', a small corner shop in Sunderland set up by 7 women. It all began in 1975. Margaret Elliot, whose brain-child it was, explains. 'Pete worked in a building co-operative, and one day one of the men involved came round to supper. He told me how co-operatives worked and straight away I thought, "That's it — that's the way it *should* be. Shared work, shared responsibilities, shared profits. Not bosses and workers, but everyone working for each other".'

Margaret's idea was that she and her friends could get together and set up something which would bring them in some income, get them out of the house, enable them to meet people, and generally enable them to share more in each other's lives and problems. A co-operative corner shop where they could also take it in turns to look after the children seemed like the best idea.

The encouraging thing is that when she had this idea she had no money at all, no clue how to raise it, and no business experience or know-how. She had left school at 15 and gone straight into a job as a clerk at a local catalogue firm. Yet she and her friends succeeded.

The 'Kennington Office Cleaners Co-operative' was set up in 1978 by a number of women who all worked as part-time office cleaners, and all of whom were dissatisfied with the work. Once they had realised that they might be better off working for themselves, they started to plan how they could do so, and chose to form themselves into a co-operative. It was quite hard to get going, but once they got their first contract other work started to come in by word of mouth and personal recommendation. Their high quality service ensures that this will continue.

In 1980 there were 19 women members, all working part-time between 6 and 15 hours a week. The company had a projected turnover for 1980 of £20,000, and held eight contracts. (Details — *In the Making*, see below.)

- The best single book about co-operatives is *Workers' Co-operatives — A Handbook*, by Cockerton, Gilmour-White, Pearce and Whyatt (Aberdeen People's Press, 163 King Street, Aberdeen, 1980, £2.25).
- *In the Making* is a valuable annual directory of co-operative projects, including articles, advice, etc. (£1.50 from 44 Albion Road, Sutton, Surrey.)
- *The Industrial Common Ownership Movement* (ICOM), Beechwood College, Elmete Lane, Roundhay, Leeds LS2 2LQ gives help and advice to co-operatives. It keeps a 'register of skills' of people who have experience in forming and developing workers' co-operatives who are willing to come and give advice. They also run useful short weekend courses. Write for

details.
- *People Power — Community and Work Groups in Action*, by Tony Gibson (Penguin, 1979, £1.25) contains useful case-studies of groups who have set up small co-operatives.

Dreaming of a business.

Chapter 10: Money

INTRODUCTION — SIGNING ON — WAYS OF MAKING YOUR MONEY GO
FURTHER — PROBLEMS WITH BENEFIT

This may be the first chapter that you have turned to, if your biggest single
problem about being unemployed s simply the money.
 So, a word about what this chapter does and doesn't give you:
 • It DOES explain briefly how to sign on and claim benefit, both Unemploy-
   ment Benefit and Supplementary Benefit (p. 113).
 • It DOES show ways in which you could either save money or increase
   your income (with a checklist for each benefit system) (p. 115).
 • It DOES give you a complete stage-by-stage budgeting exercise, so that
   you can work out where you can make savings (p. 121)
 • It DOES give you advice on some of the problems that can arise for people
   receiving benefits (p. 123).
 • It DOES tell you where to go for further help and advice (p. 124).
 • It DOES NOT spell out the detailed workings of the two benefit systems.
   We realise that in some situations, your only way round a particular prob-
   lem may be to have this detailed information — so we DO tell you where
   to go to get it (p. 124).

The chapter is divided into 3 sections:
 • Are you out of work, but not yet signed on?
                                                    GO TO SECTION 1.
 • Have you already signed on? There may be ways in which you can make
   your money go further.
                                                    GO TO SECTION 2.
 • Have you met any problems with signing on or claiming benefit?
                                                    GO TO SECTION 3.

SECTION 1: SIGNING ON

1.   At the Jobcentre

a)   As soon as you become unemployed, you should register for work at once,
      at a Jobcentre (or, if you are under 19, at a Careers Office). They will then
      give you a form to take to the nearest Unemployment Benefit Office.

2.   At the Unemployment Benefit Office

a)   Go to the Unemployment Benefit Office and find the 'fresh claims' desk.
      Tell them that you are unemployed and looking for work, and want to
      claim benefit. They will ask you a few questions (e.g. whether you have

any family, and who your last employer was) and will tell you when your signing-on time will be each fortnight.

b)   *If you have not worked before* you will not be able to claim any Unemployment Benefit, so you must claim for Supplementary Benefit. (You will actually receive a letter in two weeks time, telling you that you do not qualify for Unemployment Benefit. Don't worry — this doesn't make any difference to your claim for Supplementary Benefit.) Ask for a form B1 to take to the Social Security Office.

c)   *If you have worked before*, take either your P45 from your last job or a note of your last National Insurance number with you.

   (i)   If you are *certain* that you will receive full Unemployment Benefit, then you don't need to do anything more; your first Giro will arrive in about two weeks.

   (ii) If you are *not certain* that you will get full Unemployment Benefit, either because you have not been working for long, or because you did not work all the time for the last two years, then you *may* either receive no Unemployment Benefit, or less than the full rate. You should therefore also apply for Supplementary Benefit, by asking for a form B1.

   (iii) If you think you will get full benefit, but will be very hard up for the next two weeks until your first Giro arrives, you should also apply for Supplementary Benefit.

3.   At the Social Security Office

If you are claiming Supplementary Benefit, you should now either visit your nearest office of the DHSS (Department of Health and Social Security), or phone up to arrange an appointment for an interviewer. Be on time, or you'll have to wait for hours. At your interview the DHSS will work out how much money you need according to their rules and rates which are fixed by law (and updated every November). You should take with you to the DHSS:

- Your National Insurance number.
- Your last pay slip.
- Your rent book or mortgage details.

If you have the following, you should also take them:

- Your last electricity/gas bill, water bill, and evidence of rates.
- Any savings book or statement from your bank, post office or building society.
- Your child benefit (family allowance) book.
- (Men) Details of anything your wife may be earning.
- Details of any benefits you are receiving.
- Details of any HP agreements.
- Something to read, or some knitting, for while you wait!

If you have capital or savings over £2,000, you will not be able to receive any Supplementary Benefit. This includes cash, money in the bank, redundancy payments, and even the surrender value of a life insurance or endowment policy. (You can still claim for a rent and rate rebate, and free welfare benefits, but otherwise, you are on your own.)

You can take a friend with you to your interview; this often gives people a lot more confidence.

At the interview, you will be asked about your income, and then about your needs. You should also be asked if you want to claim for any 'Additional Requirements', for a list of various things — see below, p. 119. If they do not ask, you must mention them yourself, if you want extra benefit for them.

The interviewing officer will write down everything you say, and then ask you to sign it, as a statement. Read it carefully to be sure that it is correct, and then sign.

If your fares to the DHSS come to more than £1, you should claim these back. Ask the interviewing officer.

4.   Your benefit arrives

Unemployment Benefit: Your benefit will be made up of a basic rate, plus an addition for any dependants, plus a possible addition for Earnings Related Benefit. (This is being phased out during 1981, and will be gone by June 1982.) Your Unemployment Benefit will stop after a year, after which you should claim for Supplementary Benefit. Earnings Related Benefit runs out after 6 months. It comes by Giro every two weeks. Check, every now and then, whether you are entitled to Supplementary Benefit — e.g. every November and when Earnings Related Benefit runs out.

Supplementary Benefit: Your benefit will be made up of a basic allowance for yourself, or for yourself and your partner, *plus* an addition for any dependants you have, *plus* the cost of your rent and rates or the interest on your mortgage (not the capital repayments) for the fortnight, *plus* any increase for Additional Requirements.

There should be a note with your first Giro, called a 'Notice of Assessment', explaining how it has been worked out (Form A 124). If there is not, simply phone the DHSS and ask them to send you one. If you have any problems over your benefit, see below, p. 123. Your benefit comes by Giro every two weeks.

5.   From now on

If you live more than six miles from the nearest Unemployment Benefit Office, you can sign on postally. Otherwise, you sign on every two weeks, unless you have some income to declare, in which case they like you to sign on weekly, and to give full details of your earnings. If you have any problems, consult Section 3, below.

SECTION 2: WAYS OF MAKING YOUR MONEY GO FURTHER

If you are on Unemployment Benefit:

	Tick if yes	If no, see number below
Are you sure you are getting the correct benefit?	☐	No. 1
Have you applied for a tax rebate?	☐	No. 2
Have you applied for a rent and rate rebate and allowance?	☐	No. 3

	Tick if yes	If no, see number below
Are you earning an extra £4.50 a week or more?	☐	No. 4
(Men) Is your partner earning?	☐	No. 5
Have you considered taking a lodger?	☐	No. 6
Have you applied for free prescriptions, etc?	☐	No. 7
Have you applied for free school meals for your children?	☐	No. 8
Have you asked to have your mortgage or HP repayments suspended?	☐	No. 9
Are you paying your fuel bills, etc. in the easiest way?	☐	No. 10
You may be eligible for Supplementary Benefit. Have you applied?	☐	No. 11
Have you drawn up a budget of your income and expenditure?	☐	No. 16

If you are on Supplementary Benefit:

	Tick if yes	If no, see number below
Are you sure you are getting the correct benefit?	☐	No. 12
Are you receiving any 'Additional Requirements' you might be entitled to?	☐	No. 13
Have you applied for a tax rebate?	☐	No. 2
Are you earning an extra £4 a week?	☐	No. 14
Is your partner earning an extra £4 a week?	☐	No. 14
Have you considered taking in a lodger?	☐	No. 6
Have you applied for free school meals, free prescriptions, etc. that you are entitled to automatically?	☐	Nos. 7, 8
Have you asked to have your mortgage or HP repayments suspended?	☐	No. 9
Have you enquired about Single Payments?	☐	No. 15
Are you paying your fuel bills, etc. in the easiest way?	☐	No. 10
Have you drawn up a budget of your income and expenditure?	☐	No. 16

1. Are you sure you are getting the correct (Unemployment) Benefit?

Phone or visit the Unemployment Benefit Office, and ask them to explain, or check. If you are still unsatisfied, ask an advice centre for help (see p.124). You can also get help from the Claimants' Union. (See p. 125).

2. Have you applied for a tax rebate?

If you have paid any PAYE income tax since last April 5th, you ought to get some back. After being out of work for four weeks, you can claim by asking for a form P50 from the Tax Office (under 'Inland Revenue' in the phone book), or from the UBO. After April 1982 things will change, since benefits will become taxable. This will mean that tax rebates will be far smaller, or that you may not get one at all.

3. Have you put in a claim for a rent and rate rebate or allowance?

These are available to anyone on a low or lowish income; it is best to apply after you have tried for Supplementary Benefit. The easiest way to find out if you qualify is to apply. Phone the local council and ask for the rebates department, and ask them to post you the leaflets and forms about it. Then fill them in and post them. (If you want help with filling in forms, see p. 128.)

One word, however: if you are a private tenant, the council will not necessarily take your *actual* rent into account when they work out if you qualify. They will only give you one on the basis of what they decide is a 'fair rent' for your house. If this happens, see p. viii.

This may annoy you. What you need to do is contact the Rent Officer (phone book) to get his advice about having your house or flat registered officially, and the rent fixed. The Citizens Advice Bureau can help here.

4. Are you earning an extra £4.50 a week, or more?

Here are the rules about earning money while on Unemployment Benefit:
- You can earn as much as you like on Sunday. You are only paid benefit to cover you for Monday — Saturday. Possibilities? See No. 14 below.
- You can earn up to 75p a day, or £4.50 a week (after expense), without losing any benefit. (This is the 1981 figure. It may change.)
  Expenses: this covers anything you have to spend in order to earn your money — child-minding fees, playgroup fees, overalls, materials, bus-fares, special cleaning, and any other reasonable expense. You can also deduct tax and insurance.
- You can choose to give up the benefit for one or two days, if you find part-time work for that day. If you can earn more than one sixth of your weekly benefit, on any one day, it is obviously worth your while to do so.

However (you're right, there had to be some snags!), the rules do state that if you find work as an employee (as against being self-employed), the work must not be in your normal line of work.

The rules also state that it is not on for you to establish yourself with a relatively well-paid regular job for 2 or 3 days a week, and then claim benefit for the other days.

You must still be available for full-time work.

5. The rules about dependants earning money

A man receives an increase of benefit for his wife, or partner (see No. 6 for the conditions), and she can earn up to that amount without any problems. If she earns over that amount, the increase is stopped altogether.

She can alternatively choose to work, and earn over the rate of increase, which has no effect on the man's benefit.

A woman claiming benefit cannot receive any increase for a dependant male (even if he is looking after the children) unless he is incapable of work — and will remain so for 6 months. By 'work', they mean 'earning up to the Supplementary Benefit level for a single person'. Presumably he could still earn a little, without invoking Departmental Wrath. See the *Rights Guide* mentioned on p. 124 for the full story.

6. Have you considered taking in a lodger?

If you were to take in a lodger, this would be counted as being the work of the woman of the house, and so only some will be counted as income. Bear in mind that any income you receive will affect the amount of rent and rate rebates you can receive.

7. Have you applied for free prescriptions, and other welfare benefits?

With prescriptions at £1 a throw, life can get very expensive if you or your children start to fall to bits. People on very low incomes and everyone on Supplementary Benefit, may receive free prescriptions, dental treatment, glasses, and milk and vitamins for under-fives. There is a leaflet which explains whether you might qualify for any of these items, with the form to fill in. It is available in post offices (or by phoning the DHSS). (See p.128 if you want help with forms.)

You can also get help with fares to hospital if (and only if) the cost of the fares would reduce your income so that it fell within the Supplementary Benefit levels. One way to find out if it does is to apply using form H11 (available at hospitals) or by phoning the DHSS.

You may also be able to claim Local Authority benefits such as school uniform, clothing allowances, travel and holiday grants. Ask the Citizens Advice Bureau.

8. Have you applied for your children to get free school meals?

Every county and city council has slightly different rules about who can or can't get free meals. The simplest way to find out if you qualify or not is to apply. Phone your Area Education Office (see p. vii) and ask for an application form. You may have to persist. People receiving Supplementary Benefit or Family Income Supplement ought to get free school meals automatically.

9. Have you asked to have your mortgage or HP capital repayments suspended?

Your Building Society and any firms you have HP agreements with may agree to suspend capital repayments until you can find a job again. Write to them, explaining your position. It may also be worth writing to the DHSS.

10.   Are you paying your fuel bills, etc. in the easiest way?

- You can buy stamps for your TV licence and your phone bill at the post office.
- You can buy stamps for your gas and electricity bills at the Gas and Electricity Showrooms, and also quite often at local shops, or at the post office. Ask if they sell them, and persuade them to if they do not.
- You can pay an agreed amount each week or month towards your fuel bills at the post office or by Bankers Order.
- You can have a slot meter installed if there are no other ways to meet your bills and there is real hardship.

Are you under threat of disconnection? See p. 130 for advice.

11.   You may be eligible for Supplementary Benefit

With Unemployment Benefit being reduced against inflation by 5% each year, and with Earnings Related Supplement disappearing altogether, more and more people on Unemployment Benefit will find that they receive less from Unemployment Benefit than they would under Supplementary Benefit. To find out if you qualify, simply ask for a form B1 at the UBO to take or send to the DHSS, or fill in one of the forms at the post office.

If you do receive Supplementary Benefit, a few things will change. You will automatically qualify for free school meals, free prescriptions and welfare benefits. On the other hand, you will fall within the Supplementary Benefit earning rules, which are a lot tighter than the Unemployment Benefit rules. So if you can earn any extra part-time cash, this may be a better solution. Once you are on Supplementary Benefit, even if it is only £1 a week, you fall under all their rules.

12.   Are you sure you are getting the correct (Supplementary) Benefit?

If you are not sure, phone the DHSS and ask them to explain; and/or ask them to send you a form A124 (Notice of Assessment); and/or consult the Citizens Advice Bureau, for their help (see below, p. viii); and/or you might want to make an appeal against the DHSS's decision about your benefit — see below, p. 127.

13.   Are you receiving any 'Additional Requirements' benefit increase that you might be entitled to?

You may be entitled to a regular fortnightly addition, on top of your normal benefit, if you meet one or more of these conditions:

Additions for heating if:
- You suffer chronic ill-health or serious illness.
- Your home is difficult to heat.

- You have central heating.
- You live on a housing estate with an excessively expensive heating system.
- You are being charged for heating, hot water and lighting in your rent, and the charges are above the DHSS's figures for the weekly cost of these.
- A member of your household is under 5 or over 70.
- You are on a mobility or attendance allowance.

Other additions:
- If you need more than one bath a week for medical reasons.
- Blindness or recent blindness.
- Special diet (various conditions).
- Domestic assistance necessary (various conditions).
- *Hire purchase costs* (various conditions).
- Laundry/launderette (various conditions).
- Special wear and tear on clothing due to mental or physical problems.
- Furniture storage (various conditions).
- Regular fares to hospital (various conditions).

If you think you might qualify: ask the DHSS, and/or get advice (see p. viii below).

14.   Are you earning an extra £4 a week, and is your partner earning an extra £4 a week?

You can (both) earn up to £4 a week (after expenses) without losing any bene-fit. Anything earned over £4 must be declared, and your benefit will be reduced accordingly. (This is the 1981 figure. It may change.) Even if you earn less than £4 a week, you should declare it.
Expenses — see No. 4. The same rules apply.

Ideas? Gardening, casual work, car/bike/gadget/building repairs, odd-jobbing, delivering leaflets, freelancing, window-cleaning, 'anything legal considered', making things for sale by mail-order, on a market stall, etc. See *Earning Money at Home* (Consumers Association, £2.95) which is full of ideas (should be in the library).

15.   Single Payments

These are 'one-off' payments which can be paid to people on Supplementary Benefit who need various essential items. To get them you must have less than £300 in savings. (Any savings that you have above £300 will be taken as contri-buting towards the cost of the item.)
Single Payments can be made to meet the costs of:
- A new-born baby.
- Hospital visiting.
- Bedding, to replace worn out sheets, blankets, etc. or to buy them if you don't have any.
- Furniture, such as curtains, floor covering, tables, chairs, bed and mattres-

ses, cupboards, cooker, iron, heaters. You will be expected to buy second-hand wherever possible (except mattresses). You can claim for a garden fork, spade or shears, if you have a garden (or allotment).

- Costs involved in moving house (in some cases), and household redecoration and repairs.
- A returnable deposit that you may need to pay before you can start renting a house or flat.
- Fuel debts.
- Expenses on starting work.
- HP debts — in certain circumstances.
- Clothing and footwear, but only in certain special circumstances (and not for normal wear and tear): for a pregnant woman, for the birth of a child, or for rapid weight loss or gain; for heavy wear and tear caused by disability; for accidental loss, damage or destruction; special clothing items needed for a physical or mental illness or disability; for admittance to hospital.
- Funeral expenses.

How do you apply? Write to the DHSS, simply requesting a Single Payment to cover the cost of whatever it is. State the cost and your circumstances, and ask their advice.

The *National Welfare Benefits Handbook* (see p. 124) is an excellent and recommended source of guidance on this and all Supplementary Benefits problems.

16. Have you drawn up a complete budget of your income and expenditure, and planned where you could make savings?

Budgeting is never fun, but if you are on a low income, it is essential. Here are some guidelines.

Making up a fortnightly budget sheet (7 steps)

Step 1. Make a complete list of all your fortnightly income.. . . . (Total 1)
Step 2. Make a list of all your regular financial commitments (bills, etc. not food, etc.).

(13 times table)

13	26	39	52	65	78	91	104	117	130	143	156	169	182	195	208
1	2	3	4	5	6	7	8	9	10	11	12	13	14	15	16

- *Electricity bill:* multiply the last quarter's bill by 2, and then divide by 13. Remember that winter bills are double summer bills.   Total:. . . . .

- *Gas bill:* ditto.   Total:. . . . .

- *Telephone bill:* ditto   Total:. . . . .

- *Coal, Calor Gas:* needs its own special calculation.   Total:. . . . .

- *Weekly rent and rates:* multiply by 2.   Total:. . . . .

- *Mortgage repayments:* multiply your monthly repayment by 6, and then divide by 13.   Total:. . . . .

- *Children's school lunches* for 10 days.   Total:. . . . .

- *Children's weekly pocket money*, multiplied by 2.   Total:. . . . .

- *TV rent and licence*   Total:. . . . .

- *Hire purchase*   Total:. . . . .

Regular commitments total

. . . .
(Total 2)

Step 3. Car or Motor-bike. Work out the annual costs of your
- Road tax: . . . .
- Insurance: . . . .
- MoT + services: . . . .
- AA/RAC membership: . . . .
- Extra repairs: . . . .
                    TOTAL:. . . . .

Divide by 2 and then by 13, to obtain
cost per fortnight:. . . . .

Petrol, oil per week:. . . . . . .   Cost per fortnight:. . . . .

Total cost of running the car/bike per fortnight:. . . . .

(Total 3)

Step 4. For two weeks, keep a check on all your household expenditure. Make several separate lists, headed food, 'things', papers, bus/train/tube, drink, tobacco, stamps, miscellaneous, cost of being constructively unemployed (evening classes, materials, photocopying, etc.)

Total cost for 2 weeks:. . . . .
(Total 4)

Step 5. Make an estimate of the cost of any essential expenditure that is coming up over the next three months, e.g. on clothes, shoes, presents, repairs, etc.

Total:. . . . .

Multiply by 2 and divide by 13. Cost per fortnight:. . . . .
(Total 5)

Step 6. Add together totals 2 to 5:

TOTAL FORTNIGHTLY EXPENDITURE:. . . . .

TOTAL FORTNIGHTLY INCOME:. . . . .

FORTNIGHTLY PROBLEM :

You will probably now need a glass of brandy/week in hospital/hole in the ground to bury yourself in/false beard and passport/new job, as fast as you can
If (sadly) none of these seem to be realistic solutions to your problems, you have to start making some difficult decisions.

Step 7. Where can you reduce your expenditure? What gets cut out? And, can you increase your income or reduce your expenditure by any of the other measures suggested in this chapter?

Saving money. It is always difficult to recommend ways to save money — everyone is so different, and we all have our different ideas of what is really important, and what can perhaps go. If you belong to a group of people who are also out of work, as outlined in Chapter 5, a general discussion of ways in which you save money would probably yield more ideas than any list printed here would. So just one money-saving idea: form a food co-operative with other local people, and buy your food at wholesale prices, sharing the work out between you. (See Chapter 8.)

SECTION 3. PROBLEMS WITH BENEFIT

This Handbook has tried to set out the benefit systems as clearly as possible. However, they can be quite complicated, and of course you can run into problems. Here are some of the commoner problems — in particular, check No. 1, which lists other sources of help. You're *not alone* with any problems you may have; there are other people and places who will help.

Do you have any of these difficulties?	*Tick if yes*	*If yes, see page*
1. You want somebody to help with benefit problems.	☐	124
2. You have had your Unemployment Benefit cut, or your Supplementary Benefit reduced for six weeks, under the Industrial Misconduct Rule.	☐	125

		Tick if yes	If yes, see page
3.	You disagree with a decision about your benefit and want to appeal.	☐	126
4.	You would like help with forms.	☐	128
5.	Your Giro hasn't arrived.	☐	128
6.	Your benefit has been cut under the cohabitation rule.	☐	129
7.	The Unemployment Review Officer wants to see you.	☐	129
8.	You want to go on holiday while on benefit.	☐	129
9.	You want to do voluntary work while on benefit.	☐	129
10.	You are not getting enough benefit for the rent.	☐	129
11.	You are threatened with disconnection.	☐	130
12.	You are starting work.	☐	130

1.   Sources of help and advice

a) Try to get a clear explanation from the UBO (Unemployment Benefit Office) or the DHSS (Department of Health and Social Security) by phoning up, and asking to speak to somebody who can explain your muddle.

b) There are two *excellent* guide-books which explain the ins and outs of all the benefit systems far more thoroughly than this chapter can ever hope to do. If you can buy these, or ask your library to buy them, it will be well worthwhile:

Rights Guide to Non-Means-Tested Social Security Benefits (£1.35). This covers Unemployment Benefit (also sickness benefit, child benefit and benefits for the disabled).

National Welfare Benefits Handbook (£1.35). This covers Supplementary Benefit (also Family Income Supplement, and housing and welfare benefits).

They are both available from the Child Poverty Action Group, 1 Macklin Street, London WC2B 5NH. If you want thorough reliable information on benefits, this is the place to find it.

c) The *Citizens Advice Bureau* (p.viii) can give you help and advice, and sometimes provide somebody to accompany you to a Tribunal, for an Appeal. They know the system, and are keen to help.

d) *Local advice centres and groups.* There are various excellent groups scattered around the country, which can offer you help. It is not so easy to tell you how to find these, however, but if you ask at the Citizens Advice Bureau, look in the phone book or ask around for one of the following,

you ought to strike lucky:
- Neighbourhood Advice Centres. Neighbourhood Law Centres, etc.
- *Claimants' Union branches.* The Claimants' Union has groups all over the country. They work on a mutual self-help basis, with claimants giving each other support with problems, at interviews and at appeals. They have found from long experience that if you are prepared to fight for your rights, you will get them, and that it is always easier to do this if you have support from someone sympathetic. Write to the Federation of Claimants' Unions, Dame Colet House, Ben Jonson Road; Stepney Green, London E1, enclosing a stamp, for an up-to-date list of local branches.
- *Child Poverty Action Groups,* which campaign for better rights for all people on low incomes. They are a source of sound advice and experience. Write to them at 1 Macklin Street, London WC2B 5NH for a list of groups.

Finally, if you cannot solve your problem locally, or if you want to make an appeal against an Appeal Tribunal decision, you can write to the Citizens' Rights Office, 1 Macklin Street, London WC2B 5HN, 01-242 6672.
- The local library may also be able to help with addresses, etc.

2. The industrial misconduct rule
Under the rule, a person can be disqualified from benefit for up to six weeks, if he or she lost the last job through misconduct (i.e. was dismissed), or left the job 'voluntarily, without just cause'.

The purpose of the legislation (one can only assume) is to penalise people for making free choices about where they work, and to penalise them for whatever activities cause them to be dismissed from their work. Or, to put it another way, the legislation is to encourage people to be docile and subservient employees. It is really a bit of antiquated (and vindictive) nonsense that ought to be abolished.

When you first sign on, the UBO writes a letter to your previous employer, asking why you left the job. If the reason seems to be that you either left through misconduct, or 'voluntarily, without good cause', your unemployment benefit will be suspended for up to six weeks. You may be told that this is happening, but alternatively, you may only learn about it when you get around to asking what has happened to your benefit. What can you do?

a) *Do you argue with the ruling?*
(i) 'Voluntarily, without good cause'. The following would be grounds for an appeal:
- Reasons connected with housing.
- Excessive travel costs.
- Dangerous or unhealthy conditions.
- An excessive burden on your marriage or relationship.

- Physical difficulty in doing the job.
- You left to start another job, which fell through, through no fault of your own.
- You were just trying the job out, to see if it suited you, and you found that it didn't.

(ii)   *You were dismissed.* If your employer tells the UBO that you were sacked for misconduct, the UBO will send you a copy of his letter, and ask you to comment on it. The Insurance Officer has to decide whether you should be disqualified or not, so it is up to you to explain your side of things as fully as you can. In a recent ruling, the Commissioners ruled that 'misconduct' must mean something 'blameworthy, reprehensible or wrong'. If the Insurance Officer decides that the blame should be shared, he can reduce the period of your disqualification.

(b)   *You can appeal*

If your reply does not cause your disqualification to be over-ruled or reduced, and you believe that you were in the right, you can appeal within 28 days to the National Insurance local tribunal. Ask at the UBO.

(c)   *You can apply for Supplementary Benefit*

Your B1 will inform the DHSS about the situation, and if you qualify for benefit, your personal benefit (but not the other components of your benefit) will be reduced by 40% for up to 6 weeks. This will set you back by about £50 over the 6 weeks. You can appeal against this reduction, this time to the Supplementary Benefits Appeal Tribunal, again within 28 days.

(d)   *Will the reduction cause you hardship?*

You can also appeal to have the reduction reduced by half so long as you do not have more than £100 capital savings, *and* if one of the following applies:

- You or one of your dependants is pregnant, or seriously ill.
- Your family includes a child under five.
- Your previous job only lasted six weeks or less.
- Your average earnings over the last six weeks were less than you would be getting for yourself and family if you were on full Supplementary Benefit, plus £4.
- You are not receiving benefit for your full rent.
- You have repayments on your mortgage or HP.
- You can show similar reasons, and the reduction is going to cause you hardship.

NOTE: If you do not receive any Supplementary Benefit, you can still claim for a rent and rate rebate or allowance. See p. 117.

More information: All that exists is Leaflet UBL 48 *Suspension of Unemployment Benefit,* which is a very inadequate guide to the rule. *Rights Guide to Non-Means-Tested Social Security Benefits* (See p. 124) carries several pages of good detailed advice, and is far more use.

3. You disagree with a decision about your benefit, and want to appeal

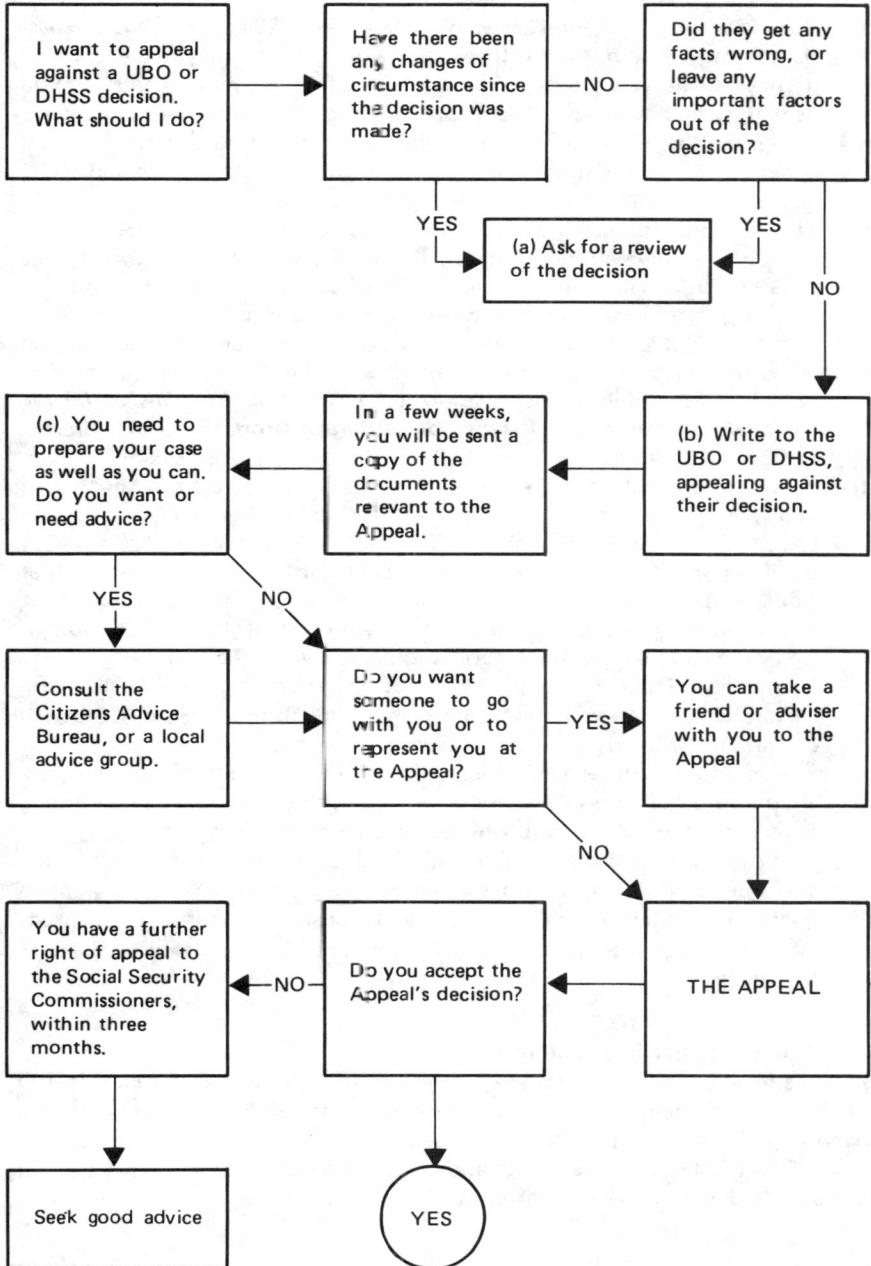

```
┌─────────────────┐      ┌─────────────────┐      ┌─────────────────┐
│ I want to appeal│      │ Have there been │      │ Did they get any│
│ against a UBO or│ ───▶ │ any changes of  │      │ facts wrong, or │
│ DHSS decision.  │      │ circumstance    │─NO─  │ leave any       │
│ What should I do?│     │ since the       │      │ important factors│
│                 │      │ decision was    │      │ out of the      │
│                 │      │ made?           │      │ decision?       │
└─────────────────┘      └─────────────────┘      └─────────────────┘
                              │ YES                     │ YES
                              ▼                          ▼
                         ┌──────────────────────────────┐
                         │ (a) Ask for a review          │
                         │ of the decision               │
                         └──────────────────────────────┘
                                                    │ NO
                                                    ▼
```

```
┌─────────────────┐      ┌─────────────────┐      ┌─────────────────┐
│ (c) You need to │      │ In a few weeks, │      │ (b) Write to the│
│ prepare your    │ ◀──  │ you will be sent│ ◀──  │ UBO or DHSS,    │
│ case as well as │      │ a copy of the   │      │ appealing against│
│ you can. Do you │      │ documents       │      │ their decision. │
│ want or need    │      │ relevant to the │      │                 │
│ advice?         │      │ Appeal.         │      │                 │
└─────────────────┘      └─────────────────┘      └─────────────────┘
   │ YES        │ NO
   ▼             \
┌─────────────────┐      ┌─────────────────┐      ┌─────────────────┐
│ Consult the     │      │ Do you want     │      │ You can take a  │
│ Citizens Advice │ ───▶ │ someone to go   │─YES─▶│ friend or adviser│
│ Bureau, or a    │      │ with you or to  │      │ with you to the │
│ local advice    │      │ represent you at│      │ Appeal          │
│ group.          │      │ the Appeal?     │      │                 │
└─────────────────┘      └─────────────────┘      └─────────────────┘
                                   │ NO                    │
                                   ▼                       ▼
┌─────────────────┐      ┌─────────────────┐      ┌─────────────────┐
│ You have a      │      │ Do you accept   │      │                 │
│ further right of│      │ the Appeal's    │      │ THE APPEAL      │
│ appeal to the   │ ◀NO─ │ decision?       │ ◀──  │                 │
│ Social Security │      │                 │      │                 │
│ Commissioners,  │      │                 │      │                 │
│ within three    │      │                 │      │                 │
│ months.         │      │                 │      │                 │
└─────────────────┘      └─────────────────┘      └─────────────────┘
        │                       │
        ▼                       ▼
┌─────────────────┐          ╭───────╮
│ Seek good advice│          │  YES  │
└─────────────────┘          ╰───────╯
```

(a)   It is sometimes easier to have a decision changed or corrected in this manner, than by going all through appeal.

(b)   You have to send in your letter of appeal within 28 days. (If you are past the 28 day limit, you can still appeal, if you explain why you are late, and if they decide you have 'good cause' for being late.) At some stage you need to write down why you are appealing in detail, giving your arguments. You can either do it now, or you can just send in a quick letter saying 'I appeal against the decision of such-and-such date, about such-and-such', and then take your time to prepare a good letter later.

(c)   The Appeals Tribunal is fairly informal, but the members of the Tribunal can only make a proper decision if they know all the arguments. The DHSS or UBO will be arguing their point of view against you, so it is up to you to prepare your own case as carefully as you can. The Citizens Advice Bureau have a lot of experience at this kind of thing, and can help you. So can other advice and support groups. (See p. viii) If you need more information, take a look at *Social Security Appeals, a guide published by the National Association of Citizens Advice Bureau* (from the CAB, libraries, or £2 from bookshops or from NACAB, 110 Drury Lane, London WC2).

(d)   It is always a help to take someone with you for advice and support. If you want to be represented, instead of conducting your case yourself, you can arrange this too, through the Citizens Advice Bureau, or through your trade union. You can also ask someone to attend as a witness, if that would help your case (e.g. a community worker, to state that you did not receive any earnings for local community work you might have been doing). Everyone you take along will receive expenses, including you.

(e)   *The Appeal.* There are no strict rules of procedure. You will be given opportunities to state your case, to put questions to the DHSS or UBO officer, and for your adviser and witnesses to speak.

   For appeals concerning Unemployment Benefit matters, you will usually be told the Appeal's decision then and there. For appeals concerning Supplementary Benefit, you will be told two or three days later.

(f)   You can appeal against the Tribunal's decision. You should definitely get advice about doing this, so that you present a good case. You may want to question an interpretation of the law, for instance. Consult the Citizens Advice Bureau, the Child Poverty Action Group or the Claimants' Union . (See p. viii)

4.   You would like help with forms

When claiming benefits, you sometimes need to fill in a form. Don't let this stop you getting your money. If you find a form difficult, ask for help with it from a friend or colleague, or from an advice centre or Citizens Advice Bureau. (See p. viii) Don't feel embarrassed about asking — even people with university degrees have been known to hide from forms in fright!

5.   Your Giro hasn't arrived

If your Giro doesn't arrive, phone up straight away to check if it has been sent.

Ditto if you lose it. The DHSS is meant to replace it, either immediately or as soon as possible.

6. Your benefit has been cut under the cohabitation rule
Single women claimants sometimes have their benefit cut off if the DHSS get the idea that they've got a man cohabiting with them. If this happens to you, seek advice. See p. viii.

7. The Unemployment Review Officer wants to see me.
It helps to be clear on one thing: the staff at the Jobcentre have no interest in pushing you into a job that you don't want and won't like. It only rebounds on them, because they get dissatisfied employers that way, which means fewer jobs being channelled through their office.

The special 'Unemployment Review Officers' attached to the DHSS, however, *do* have an interest in getting you off the benefits system, and that is why they exist. So they will be suggesting that you think about other kinds of work, which you may not want, or which you may not be suited to. If you have been unemployed for over a year (or six months, in some areas) they may tell you that if you don't find work within three weeks, they will cut you off from benefit (especially if you are single). Their reason for doing this is because they find, quite simply, that it works, in their terms — a large number of people cut off in this way do not bother to appeal, or to apply for benefit again. If you are furious at being cut off like this, you should appeal, and explain your reasons for being angry — the things you are doing to find work, the state of the economy, etc.

8. You want to go on holiday while on benefit
Ask for a holiday form and fill it in. You will be asked to leave an address where they can contact you, should anything turn up. If you are camping, or travelling, you can arrange that you phone them up every so often, or that you leave a contact address. As the rules stand, you cannot leave the country without losing benefit.

9. What about voluntary work?
All sorts of fuss and bother here, and tales of people getting cut off. The rules state that you must be 'available for work' while receiving benefit, which is easy to prove. What sometimes happens is that a nosey neighbour sees you going off down the street, and phones the UBO or DHSS to tell them you've got a job, and the next thing is — no Giro. Get on the phone straight away, to find out what is happening, and appeal, if you are cut off. (They may simply re-instate your benefit.) Ideas are changing, and it is hard to believe that anyone could really lose an appeal decision about doing voluntary work. If you do, appeal against the Tribunal's decision. (And tell the local press!)

10. You are not being given enough (Supplementary) Benefit to cover the full rent
Where rents are high, the DHSS often apply a 'Rent Stop', and refuse to meet

the full rent. They are not allowed to do this for the first six months if you could previously afford the cost, or if it is unreasonable to ask you to move. If they do, you can appeal. You might do well to contact the Rent Officer (under 'Rent Office' in phone book), to see if you can get your rent reduced. The Citizens Advice Bureau will help you.

11.   You are threatened with disconnection

The first thing you *must* do is contact the gas or electricity board. Once you have got in touch, then they can discuss arrangements with you. Far too many people leave it all until it is too late, just hoping that the problem will go away.

There are two things that you can do:

- You can arrange a 'Fuel Direct' scheme with the DHSS, so that they take money off your Giro to cover the debt and your present consumption of fuel.
- You can arrange to make regular payments every week or month.

   You are protected from disconnection by a 'Code of Practice', which states that you should not be disconnected if:

- You have agreed to make regular payments, or to have a 'Fuel Direct' scheme.
- Your debt is for a hire purchase debt, not for actual fuel.
- There is real hardship and it is safe and practical to install a slot meter. This would then be set at a higher rate, to collect the debt.
- If you are waiting for a decision from the DHSS as to whether they can help you.

12.   Starting work

When looking for work, it is worth remembering that a job with low pay would mean that you could also claim for rent and rate rebates, and for Family Income Supplement (FIS) which is available to anyone with children who is in full-time work. You can find details about FIS in post offices, in leaflets SB1 and FIS 1.

If you were on Supplementary Benefit, you can also claim for a Single Payment, to help you with any special clothing, etc. needed for work. You can also apply to receive benefit for the first two weeks in the new job to cover you until you get paid. Ask for the form A7 at the DHSS, get your new employer to stamp it, and return it to the DHSS at once. When you start work, you can also claim fares for the first 15 days.

A final word

Many people experience great difficulty with the benefit systems. It is often true that the only way to get your rights is to struggle for them, getting help from the groups mentioned on p. 124 above.

But with unemployment here to stay, it would seem, something better *must* be developed. Local 'Consultative Councils' are needed to represent and protect the consumer — i.e. the claimant. If such bodies existed, they could lead to better mutual understanding (and even trust!) on both sides, and they could begin joint discussions about useful and long-overdue reforms. If a local action group against unemployment is set up in any area, this is one initiative that could be valuable.

Chapter 11: Time to up and go?

MOVING AWAY TO ANOTHER AREA — MOVING HOUSE — MOVING TO
LONDON — LOOKING FOR WORK ABROAD — RURAL RESETTLEMENT

MOVING AWAY TO ANOTHER AREA

If you are thinking of moving to another area, here are some suggestions.

Finding out about jobs

1. Your local Jobcentre will have details of some jobs in other areas, but not a
   complete listing. You can ask the Jobcentre to forward your name to the
   Jobcentre in the area you want to go to, or to give you the address so that
   you can write yourself. They may be able to let you know what's available.

2. Obtain copies of the local papers in that area. You can find out which they
   are by going to your nearest large library and asking for the *Advertisers'
   Annual* (which lists all the local and regional newspapers) or the Yellow Pages
   for the relevant area (under 'Newspaper and Periodical publishers'). Write to
   the head office of any local or regional papers, sending 25p to each. Ask if
   they will kindly send you a copy, preferably of an issue which carries a lot of
   job advertisements.

3. If you can obtain Yellow Pages for the area in question, you could write to
   the employment agencies listed.

Making a visit

1. The Jobsearch scheme (run by the Employment Services Division of the MSC)
   can possibly help a few people. Under this scheme, you can get a 'Temporary
   Transfer' for two weeks to look for work in another area, and you will receive
   a subsistence allowance and travel expenses, on top of your normal benefit.
   However, the decision whether to pay you or not rests with the Jobcentre at
   the other end, and they have to ask whether there are locally unemployed
   people who could do the kind of work you are looking for. If they are agree-
   able, you will be paid. You must apply before going. Details from the Jobcentre.

2. If Jobsearch cannot help you, you can still go under your own steam:
   - Go in between signing-on days, leaving a forwarding address at home, and
     you ought not to meet any problems.
   - If you need to go for longer, you can ask for a 'Holiday Form'. Under this
     arrangement, you leave a contact address (or agree to phone through regu-
     larly), and receive your benefit when you return.

- You may be asked to sign on temporarily in the place that you are going to.
- If you are young (or even if you are not), don't forget about the various ways in which you can work away from home listed in Chapter 6.

MOVING HOUSE

1. Home owners

If you own your own house, and are receiving Supplementary Benefit, you are allowed to be absent from your house for up to 6 months if you need to be, while in transition, without it being counted as capital. So your benefit will not be affected. In general, you'd do well to buy a copy of the *Rights Guide for Home Owners,* price 70p from the CPAG, I Macklin Street, London WC2 5HN.

Being unemployed, you have an opportunity to save yourself quite a sum of money if you do your own conveyancing. If you spent a full 3 weeks studying the form, learning the lingo, and doing all the bits and pieces, you might save yourself around £600 on your solicitor's bill.

Moving house.

Useful books include:
Buying and Selling a House, Marjorie Giles (National Extension College, 18 Brooklands Avenue, Cambridge, £2.95 inc. p & p) goes through the process stage by stage.
The Legal Side of Buying a House (Consumers Association, 14 Buckingham Street, London WC2N 6DS).

2. Council house tenants

If you are a council house tenant, and you want to move to a council house in another area, you have the various problems of a transfer ahead of you. Get in touch with your local housing department as a first step, for their advice. Then advertise your house in the local paper of the town or area where you want to move to, for three weeks. (See above for addresses of local papers. The *Advertisers' Annual* will also list small ad. rates.) It is really unfortunate that there is no simple system to enable people to move around yet. If there is a Neighbourhood Advice Centre of any kind in the place you want to move to, ask their help and advice. Small postcard advertisements in newsagents' windows will help, as well. A phone number will help — if you are not on the phone yourself, ask a neighbour who is, if he or she could help you.

3. Private tenants

If you are on Supplementary Benefit you may be able to claim for two rents (one in your old area and one in the new) for a short time if you *have* to pay both. (Refer to A code, Para 1842A — you may have to persist.)

As far as advice on finding a flat or house to rent goes, this is really beyond the scope of this Handbook. You need to apply the same thorough planning and efficiency that was recommended in Chapter 4 for finding a job, making sure that you cover all the possibilities — newspaper ads, agencies, estate agents, housing associations, small newsagents' notice-boards, the grapevine, asking in pubs and shops, etc. Some areas are lucky enough to have housing advice centres, which can help. Get in touch with the Citizens Advice Bureau to find out if there is one in the area you are moving to and to ask their general advice on finding a place to rent.

If you want lodgings, the social services department (in the phone book, under the name of your local authority or county council) should keep a list of good lodgings in the area you are moving to — or be able to obtain one for you. The housing department might have a list too. (But they don't usually bother much about single people.) The addresses of YMCA and YWCA hostels in England, Scotland and Wales can be obtained from the YMCA (National Office) at 2 Weymouth Street, London W1N 4AX. If you approach your own local social services department *before* you go, they might be able to find you a couple of contacts in advance. (The Jobcentre can also help you to find temporary lodgings, since they keep lists of suitable lodgings for people going on their TOPS courses.)

Leaving Home — the book called *Help!*, by Barbara Paterson (Peacock Books £1.00) is very good on all aspects of leaving home, and other problems facing young people coping with life on their own for the first time.

A few people may be able to use the Employment Transfer Scheme, which offers grant aid to people moving to another area to take up employment. However, although you can apply at your local Jobcentre, the decision whether to pay or not rests with the Jobcentre at the other end, and they only pay if there is no one on their own books who could do the kind of work you would be looking for. At a time of high unemployment, this makes the scheme fairly useless. However, you can still apply. Details from the Jobcentre.

MOVING TO LONDON

There are more jobs available in London and the South East then elsewhere, but there is also a *very severe* housing problem, and life can be very unpleasant if you don't find somewhere to live quite quickly. There is no romance in dossing on a dirty floor, really (well, not after the first few nights!).

If you are set on going, do plan carefully:

- You can get a *Directory of London Hostels* from the YMCA ([London Office], 16 Great Russell Street, London WC1B 3LR) for 50p.
- *Centrepoint Hostel* (65a Shaftesbury Avenue, London W1 [tel. 01-734 1075]) can give free accommodation to people under 25 for 3 nights. They can help people to sign on and find further accommodation and work. Doors open 8 p.m.
- *The Portobello Project* (49 Porchester Road, London W2 [01-221 4413] [Open Monday to Friday, 9-5]), will give help and assistance in finding both work and accommodation.
- *After Six* is a 24-hour telephone service offering advice and assistance, on 01-836 6534. You can reverse the charges, if you have to. You can write to them for a free booklet called *Finding a Place To Live in London* from 48 King William IV Street, London WC2.
- *Capital Radio* runs a special Jobfinder Service to help people find work. You can contact them at Capital Jobmate, on 01-636 3261.
- *Capital Appointments* (no connection to Capital Radio) offer several free lists of vacancies, covering building services, draughting, electronics, field service, mechanical engineering and production engineering. Phone 01-637 5551.

LOOKING FOR WORK ABROAD

The best source of information about jobs in Europe and overseas is the daily newspapers — the *Sun,* the *Daily Mirror* and the *Star*, etc. for skilled manual jobs, and *The Times,* the *Guardian* and the *Daily Telegraph* for skilled professional jobs.

Some of these advertisements are from agencies, so make sure the terms are right before setting off for Germany or Saudi Arabia.

Magazines such as *Construction News*, on sale in large newsagents, also carry advertisements.

Jobcentres operate a central clearing house for jobs abroad, if you ask about this.

- *Jobs and Careers Abroad* gives precise information about skills and qualifications in demand in countries overseas, both in Europe and further afield. It comes out every year, costing £5.95, from Vacation Work, Park End Street , Oxford.
- *Working Abroad: The Daily Telegraph Guide to Living and Working Overseas*, by Godfrey Golzen and Margaret Stewart (Kogan Page) gives you advice on how to get a job overseas, what kind of salary to expect, cost of living comparisons, moving house, money matters, etc. with a country-by-country survey of the job situation. £2.95 from bookshops or £3.35 from Kogan Page, 116a Pentonville Road, London N1 9JN.
- *Working in the Federal Republic of Germany,* a special booklet, should be available at the Jobcentre.

● *North Sea oil* — Jobcentres have an application scheme for all North Sea oil work. Word has it, however, that the only reliable way to get work there (unless you have a special skill that is needed) is to go up to Aberdeen, be prepared to make do for a few weeks (sleep in a van, etc.), and persistently foot your way around the offices of the oil companies every day. This way, if you are in an office when a vacancy comes up, you might be lucky. You have to be very determined to get a job. There is demand for all sorts of workers, from kitchen staff to barbers, etc. as well as for labourers.

RURAL RESETTLEMENT

It is possible that some people might want to take the opportunity to move out of the city, and into a rural area. The *Rural Resettlement Handbook* is a superbly thorough book that has been designed to help people who wish to live and work in rural areas. Coverage includes moving house (and council house swaps), looking for land, smallholdings, subsistence farming and gardening, job opportunities, self-employment, practical experiences, planning permission, finance, etc. Price £1.80 post-paid from the Rural Resettlement Group, 5 Crown Street, Oxford.

(See also the note on the *Village Action Kit* at the end of Chapter 8.)

Chapter 12: But what can we do about unemployment?

INTRODUCTION — A 12-POINT PROGRAMME TO TACKLE UNEMPLOY-
MENT — Changing our attitudes — Providing support — Changes to the benefits
system — Reclaiming the local economy — Small business initiatives — Community business ventures — Trade union initiatives — Changes within existing
firms — Schools and young people — Community energy policies — Housing and
urban and rural development — National policy — SO, WHERE DO WE START?
— RESOURCES

INTRODUCTION

With the last chapter, we come to the huge unanswered question: what can we
do about unemployment?

To write a Handbook which devoted all of its pages to what you can do about
being unemployed, as if it were simply a personal problem, would be both wrong
and deceptive. The problem is much bigger and affects our whole social, economic and political system. But what can we do?

The problem is not temporary. Unemployment has been rising since 1966. It
is not just caused by recession; that just makes things worse at present. Nor is it
confined to the UK. Most other major industrial societies are experiencing the
same problems. We are kidding ourselves if we think that a change of policy over
interest rates, taxation or import controls can really make any big difference.

Nor should we just consider the industrial countries. The Western economic
miracle has brought as much suffering to the South of the world as it has
brought prosperity to the North. Are we really seeking a simple return to life as
it was before unemployment came along? Do we really want things to be back
'as normal'? We have to think about the kind of local economic order which can
bring us a decent livelihood, without pushing the world further down its crazy
path of famine, pollution, competition for scarce resources, and the ever-present
threat of war. These issues are not separate matters.

We have further to think about the nature of the work we are all so suddenly
keen to do. It is short-sighted to simply seek to return to 'the way things were
before', as if that was somehow a perfect age. If we want to generate new jobs,
we have to ask important questions about the *quality* of those jobs. We spend so
many hours working — these hours must be a time of meaning and fulfilment,
not a time of drudgery and boredom.

It is for this kind of reason that unemployment brings with it an opportunity.
Old jobs are collapsing by the thousand every day. In the beginning, nothing re-
places them, and people are disoriented; they feel lost, and just want to have
things back the way they were. But really? No, surely not. Surely we have an
opportunity to rethink and re-organise the way we do our 'work'? We have to

switch gear, inside, so that instead of feeling negative about the loss of what we knew, we can begin to feel positive about what we can now create.

This chapter is not about inflation and interest rates and whether one political party can reduce unemployment better than another. It is about something altogether different; it is about what we can do at the local level, in our neighbourhoods and communities. The other issues are probably still important; but they are variants of the same old game, designed to get us back onto the same old path. The ideas and initiatives mentioned in this chapter are about ways in which communities can construct a new path towards a better future; an economic order rooted in our local communities, which can give us financial stability and happiness in our working lives.

The chapter is of necessity brief, and many projects that deserve to be mentioned are left out. The 12-point programme presented here is intended as a starter for discussions about unemployment; to show that there *are* many things that can be done about unemployment, and to provide a basis for local groups to build from. The chapter finishes with a listing of the addresses of relevant groups and publications, so that the initiatives mentioned here can be followed up.

A 12-POINT PROGRAMME TO TACKLE UNEMPLOYMENT

1.   Changing our attitudes

We have all grown up on the notion that if a job pays you, then it is a 'proper' job, and that if it doesn't, then it is just 'messing around'. Working in a factory making napalm bombs is thus seen as proper work, whereas doing unpaid work in a hospital is not.

A six-month period of being unemployed can be a very positive period of life, if the time is well used. It can be a time for study, for personal exploration and change, for developing new ideas about life, for involvement in local community activities. Many people decide they would rather have less money and more time, if only it were possible, precisely so as to be able to do this kind of thing. Jobsharing has been successful, and has shown that many people would welcome the chance of working for just half the week, or every other week, in order to leave time for other matters. (See p. 36.)

One escape from the 40-hour straightjacket is the self-employment option. This way of working may bring its own load of worry, but it also brings more flexibility, and greater opportunities for human fulfilment and dignity. It is perfectly possible that when we are through with this period of change, we will see some people working three days a fortnight for an employer, three days for themselves on a self-employed basis, and three days a week in the local community, perhaps instead of paying tax — why not?

The average time that a professional or executive person has to spend between jobs is currently around six months. There are something over 100,000 such people. It is a nightmare. But if firms were to give preference to new employees who had just spent six months involved in various kinds of positive activity while 'unemployed' the negative aspect of unemployment would cease overnight.

It would simply become normal to take as a sabbatical regardless of the average time between jobs. The burden of worry and attention could then be focused upon those who found themselves out of work for longer than this average span of time. It is simply a matter of a change of attitude.

2.   Providing support

Being out of paid work can bring many positive openings, as this Handbook attempts to show — but you can fee so fed up that it's hard to realise them. It is for this reason that all unemployed people ought to have support either on an individual basis, or from a self-help group, or from both, so that they can have a guiding hand through the maze of problems and opportunities that being unemployed brings. Many professional people already receive 'redundancy counselling' when they leave a job. So, why not everyone? Without this kind of help, it will be very difficult to shift the experience of being unemployed from negative to positive, so that the opportunities become real.

3.   Changes to the benefits system

There are just three changes that come to mind. If a group of unemployed people discussed how the system could be changed, they would probably come up with many more.

Firstly, there ought to be local 'Consultative Councils' (as mentioned in Chapter 10) to help claimants and benefits officers discuss their respective problems. The degree of hostility and mistrust at present existing in the system (especially between claimants and the DHSS) is appalling, and is very painful for claimants — and, presumably, for benefits officers.

Secondly, the system needs to be changed so that claimants keep the freedom to earn money.

This change could happen through a system whereby a person's *entitlement* to benefit is *increased*, the more money you earn. This has a marvellous paradoxical logic to it, which makes sense as soon as you put figures to it.

Benefit entitlement is increased by 50% of the amount that a person earns. (That's the basic rule.) Thus a person on a basic benefit of £40 per week who earned £20 of her own accord while 'unemployed' would have her benefit entitlement increased by £10 to £50 a week. Having earned £20 herself, she would receive £30 in benefit. If she earned £40, her entitlement would increase to £60, and she would receive £20 in benefit. When she earned £80, she would become free of the benefit system altogether. In this way, people could work their way towards self-employment, or towards the setting up of a small co-operative, while unemployed; an impossibility, under the present system.

Thirdly, there needs to be a new category of people, the 'voluntary unemployed', consisting of people who are busily engaged in study, personal work or community activities, who would actually prefer to remain doing what they are doing rather than take one of the scarce jobs that they would find less fulfilling. When there are not enough full-time jobs to go round, we ought not to be harassing these people to take paying jobs — we ought to be praising them. The new category might be open to anyone who could convince a panel of local people (including claimants, community workers, etc.) that they were usefully occupied for a minimum of 20 hours a week or so. They would then be granted the slightly higher long-term rate of benefit for a period of six months, and their names would be removed from the unemployment register. This would begin to open up a new option for many people, of all ages — the only condition being that they use the time constructively.

4. Reclaiming the local economy

Some words of introduction: the experience of unemployment is one of being powerless. Something has happened (the loss of the job) that is beyond our control, on both an individual and a local level.

Somewhere in the past, we lost our sense of involvement in our own local economies, and with that, our power to create our own work. The majority came to depend on a small minority to provide them with jobs. Gradually, the state came to play a role too — but an equally distant role. Jobs fell from heaven — and when they stopped falling, you just had to wait on a street corner until they started falling again.

Well, all that is changing. Government is pulling back from its fairy godmother role. Firms are increasingly tied up with multi-national conglomerates, which care not a hoot about what happens in any particular locality, unless it happens to show profits in the corporate boardrooms of Chicago or Tokyo. Not so long ago, local employers really were local. Now, who cares any more for the local economy?

Over the years, we allowed power to pass into distant hands, and those distant hands have now forgotten about us. There is only one answer to this situation — to take power back into our own hands. There is a vacuum, simply waiting for people with energy and initiative to come and fill it. This can be done in a variety of ways, all of which begin with the realisation that 'This is *our* economy — why don't we do something to help it?'

The points that follow all express this general philosophy that the fundamental way to overcome local unemployment is to reclaim influence over the local economy.

It's important to find out who owns what, who controls what, what future investment plans are, what redundancies may be likely in the future, which local firms could use a helping hand, which local people have an interest in setting up a new enterprise, which schools, colleges, chambers of commerce, rotary clubs, etc. have an interest in helping, which councillors would be willing to become involved, etc.

In carrying out this research, it ought to be possible to engage the support of social science students from local colleges, and of older school students who might be willing to do door-to-door questionnaires, etc. It might be possible to obtain a small grant from a local firm, the Council or local charity.

Research needs to be followed by activity. So read on.

5. Small business initiatives

The small business, whether it is a 20-person private firm, a seven-person co-operative, or a single self-employed person, is the life-blood of any local economy. What can localities do to help and encourage local small businesses?

The first 'Small Industries Group' was set up in West Somerset in 1977 by a group of local people who met together to ask the question, 'What can we do about unemployment?' Young people were all leaving the area in search of jobs elsewhere. The villages would die, or be taken over by retired people. They set up the group, and obtained an MSC 'STEP' grant to employ a Development Officer. For four years he and the group have been finding ways of helping local firms in difficulty, assisting people who wanted to set up new enterprises, finding premises, untangling planning problems and organising exhibitions to encourage local trade.

They are able to do work that the County Development Officer would find impossible, because he can never hope to function on such a small scale. Yet it is only on the small, local scale that people can actually know about the problems that the bakery or the printshop are having, and take appropriate steps to remedy them. There are 'Small Industries Groups' and 'Local Enterprise Trusts' which do similar work in various parts of the country, most working on tiny budgets, and each of them started up by groups of local people.

There are many other ways in which the small business community can be encouraged — local classes for would-be entrepreneurs, special workshop facilities for inventors to develop their ideas, specially designed workshop facilities where businesses can share the use of a switchboard, secretary, computer, etc. surveys to bring out potential ideas for products, etc. It is not too much to imagine that every locality might have a network to care for and nourish the local small business community. In return, the small business community will care for and nourish the locality.

6.   Community business ventures

One stage on from these initiatives is the idea that local people can actually set up, own and control their own firms, instead of depending on someone else to provide the employment.

The first such venture was a 'Community Co-operative' set up by the villagers of Llanaelhaearn on the Lleyn peninsula of North Wales. The villagers had just managed to reverse an LEA decision to close down the village school, and realised that unless they could somehow provide local employment to replace the granite quarries that had now all closed down, the village would still die, school or no school. The co-operative idea was their solution; it currently employs 15 or so people in full-time knitwear production and many more in part-time out-work. Since then, the idea has spread — to the huge urban estate of Craigmillar, on the edge of Edinburgh, for instance. There the successful Festival Society turned its attention to problems of employment in the late 1970s. Craigmillar Festival Enterprises Ltd is a multi-functional company, limited by guarantee, that takes on general building and contracting work. 'The purpose of being limited by guarantee, as opposed to having shareholders in the normal sense of a limited company, was that we wanted to be sure that the profits and the money would not go to any individuals. In terms of dividends, all the money that is made from the Company either has to be reinvested to create new jobs, or go to the Craigmillar Festival Society.'

There are now similar community-based firms in several parts of the country, and their numbers may increase, if government money is released for them under the 'Community Enterprise' programme.

There is a lot of interest in the 'Mondragon' experiment in the Basque area of Northern Spain, where workers have established a very stable and successful network of co-operatives all over the area, employing around 17,000 people. They have also set up their own social security system. This is community-based economic activity on a massive scale, but keeping the sense of locality and belonging. How could workers in Franco's fascist Spain have set up such an idealistic people's economy? The answer is that they *cared* immensely about their own locality, partly because of Franco, who did his best to wipe the Basque people off the map altogether. Once you have this basis of care, and the willingness to work together, much can be achieved.

7.   Trade union initiatives

When the same caring approach was taken up by the shop stewards of Lucas Aerospace, the result was a pioneering initiative that has evoked interest all over Europe.

Faced with redundancies resulting from cuts in defence expenditure by the Labour government in the mid '70s, the Combine Committee of Shop Stewards decided to see if they could find ways to use their skills and labour to make alternative and socially useful products instead of idling away on the dole queue.

By asking the entire workforce about possible ideas, they came up with a portfolio of around 200 possible ideas, known as the alternative Corporate Plan. The desire was there, and the product ideas, and the energy — but management never wanted to know. They just pressed ahead with the redundancies. It is a story that says a lot about community concern versus the private profit motive.

In Newcastle, the local Trades Council has set up the Centre for the Unemployed. This advice centre is also a focus for local campaigns against prejudice, and for better facilities (like free bus travel) for the unemployed. It also provides a home for the newly formed Unemployed Workers Union, which is working to help unemployed workers who are getting no regular help from their normal union. There are moves afoot to develop such centres in other parts of the country.

In the long run, we need to be breaking away from the 'ghetto' problem of the unemployed, and seeking to integrate the problems of the unemployed into the problems of the community as a whole. For this to happen, trade unions might also need to widen their roots, so as to include general community problems, instead of being focused on the workplace. If the changes foreseen in this chapter come to be, the traditional workplace may even not exist as such in the future, and unions will need to change likewise, to be able to express and represent the real needs of local people in and out of work.

8. Changes within existing firms

The problems of unemployment are inseparable from the movement for greater democracy within industry. If a job is to be worth fighting for, it must be a good one, and one way to make a job into a good one is to give more power, involvement and participation to the people who do it. Additionally, the care and concern of workers for their firm can only really develop if workers feel involved in the firm.

It is always hard to analyse the true reason why any one firm goes to the wall; but the suspicion will often remain that in many cases it has a lot to do with antiquated management practices and bad management-worker relationships. When workers feel involved, on the other hand, and where there is real trust between workers and management, they will go to extraordinary lengths to save a firm that they care about together. The future will probably see more firms going co-operative, or adopting some form of common ownership, since this is a sound basis for rewarding and pleasant shared work.

9. Schools and young people

If we want to see a growth of self-help, initiative and lively community activity, we could do worse than introduce these things into our schools. There is no reason why children should not be running their own mini-firms before the age of 16, along the lines developed by 'Young Enterprise', the CBI brainchild that teaches young people the realities of running a business. The method (which involves the setting up and running of a miniature firm, from the raising of 25p per share capital through to the actual production of whatever the group chooses) could easily be adapted for co-operative, community-based business structures.

For unemployed school-leavers, the Youth Opportunities Programme is worthy of its expansion, and could be developed into a valuable integrated system for post-school training and introduction to work. Most constructive ideas concerning community activity, basic training, voluntary work, personal development, adventure, challenge and the widening of horizons can occur within a well-designed YOP scheme, in addition to its work component, which says a lot for the programme. Furthermore, the programme allows ample scope for local people to develop their ideas and projects for young people.

10. Community energy policies

Most localities spend huge sums of money on energy every year, and very little of this money is ever seen again — it all leaves the locality. When two parishes in Pembrokeshire (Dyfed) carried out a rough 'energy audit', they found that they were spending a quarter of a million pounds a year between them on energy. They started to think — if only we could save that energy, or generate it locally, we could be using that money ourselves to help create local employment. They have established a group which is investigating possibilities, beginning with a local insulation programme.

An appropriate alteration in national energy accountancy techniques could create a system of incentives, so that localities received financial encouragement to press ahead with this kind of scheme.

Nationally, we could provide a lot of new jobs if we scrapped the whole nuclear power programme, and shifted over to a policy based on small-scale local soft-energy options, which provide more jobs for the amount of money invested.

11.   Housing and urban and rural development

It is ridiculous, and painful for those with no home to call their own, that we have such long housing waiting lists at a time of high unemployment. But with council houses at £20,000 a throw, is it any wonder? Instead of sighing and shrugging our shoulders, we should be working out how we can combine new building methods, self-help techniques, part-ownership schemes, and simple 'shell' building techniques for young couples that can allow them to expand further when they have more cash. Then we could start building houses that young people could afford.

Our cities are also so full of waste land. Planners ought to relinquish control and pass the land over to community groups, trusting local people to do their own thing. There is plenty of energy and creativity about, much of which is restricted by regulations and private interests.

We should be encouraging movement back to country areas, too, by relaxing planning procedures and encouraging rural resettlers. We are blocking so much positive, creative energy — and furthermore, paying out every week for it, through the wastefulness of unemployment benefit.

12.   National policy

Finally, then, we come to the national level. What kind of policies are appropriate here, for this transition to a new and better way of organising our working lives?

It is simple; government policy ought simply to be one of enabling, allowing and encouraging people to do what they want to do, and removing unnecessary blocks to human activity. Resources that are at present being poured down the huge drain of unemployment benefit need to be diverted to MSC area boards, to community groups and to local authorities, to allow them to tackle local unemployment in whatever ways they think best.

There is also one other way by which the government might benefit people, both employed and unemployed, and release all the funds needed to finance these initiatives: they could scrap our whole nuclear war programme.

SO, WHERE DO WE START?

In communities large and small around the country, groups of people are starting to meet together to ask themselves 'What can we do about unemployment?' In Greenwich, Newcastle, Leeds, Bradford, Torquay, Clwyd and in the Highlands and Islands, groups are working out their best strategies, building on each other's ideas and experiences.

If you want to start up a local group, to see what you can do yourselves to tackle unemployment and build something better for your community, the way to go about it is probably something like this:

- Get together with several other people who you know might think like you, who might be interested in forming such a group. Write a letter to the local paper mentioning your intention to form a group that will start to do something about unemployment, and invite readers to get in touch.
- Study the ideas in this chapter, and send for the items listed at the end of the chapter, so that you get a clear idea of the kind of initiative that might be possible.
- Call a public meeting and invite speakers who will mention some of the ideas which interest you. Arrange good publicity; leaflet the dole queues, tell the papers, radio and television. Appeal for funds to help you pay for all this by writing a letter to the local paper.
- Plan your public meeting very carefully; you could even ask the whole meeting to break up into small groups of 10, in order to give people a chance to talk over what they feel to be the best kind of approach, before returning to the full meeting.
- Use the meeting to set up a local organisation or group, and to sign up members; in this way, you will get the names and addresses of people who are interested and supportive.
- Arrange a time then and there when anyone who is interested can come along to a working meeting, to plan where you will begin. Good luck to you!

This chapter hopes to have shown some of the ways in which unemployment can bring opportunities. It has suggested what can be done to not only bring an end to the stupid and tragic waste of human life that unemployment brings, but also help us to build for ourselves a better future.

We have a dream, and we have to build on our dreams, even more so when things seem dark. The alternatives are grim, if we do not. Turning negative into positive; that is the *wei-chi* of unemployment.

RESOURCES

The Collapse of Work by Barrie Sherman and Clive Jenkins charts the future impact of microprocessors on jobs. Eyre Methuen, £3.50.

The Mighty Micro by Christopher Evans looks more widely at the changes that microprocessors will bring. Coronet books, £1.50.

Fit for Work?: analysis of youth unemployment problems and options. Colin and Mog Ball, Chameleon books, £1.95.

Local Initiatives in Great Britain (such as those listed in this chapter). A detailed review of who, how, where and why such initiatives get going, and what they do. Price £5, The Foundation for Alternatives, The Rookery, Adderbury, Banbury, Oxon OX17 3NA.

The Redistribution of Work: a short reference guide to the main issues and to many of the people and organisations involved in creating new contexts for work.

Price £1 from *Turning Point*, 9 New Road, Ironbridge, Shropshire TF8 7AU.

Turning Point Newsletter comes out twice yearly, and is excellent for keeping in touch with the new wave of changing ideas, and activities for change (not just concerning unemployment). £1 p.a. from the above address.

Community Business Ventures. Can We Make Jobs? by John Pearce, £1.50 from Local Enterprise Advisory Project, Paisley College of Technology, Paisley, or notes under the same title free from BBC, 36 CE, London W1A 1AA. Advice from Community Business Ventures Unit, 359 The Strand, London WC2R 0HP.

Lucas Aerospace and other similar Trade Union initiatives: Centre for Alternative Industrial and Technological Systems (CAITS), NELP, Longbridge Road, Dagenham, Essex RM8 2AS (s.a.e.).

Centre for the Unemployed: 5 Queen Street, Quayside, Newcastle (s.a.e.).

Unemployed Workers Union: 5 Queen Street, Quayside, Newcastle (s.a.e.).

Greenwich Action Group on Unemployment: 105 Plumstead High Street, London SE18 (s.a.e.).

Youth Opportunities Programme: for local information, contact the Careers Service (see p. 59); nationally, write for the magazine *Actions*, ed. Colin and Mog Ball, free from New Opportunity Press, 76 St James's Lane, London N10 3RD.

Rural development — The Village Action Kit, £1.95 from National Extension College, 18 Brooklands Avenue, Cambridge CB2 2HN.

Index

16 10